FLUID FILTRATION: LIQUID

Volume II

A symposium
sponsored by
ASTM Committee F-21
on Filtration and The
American Program Committee
of the Filtration Society
Philadelphia, PA, 20–22 Oct. 1986

ASTM SPECIAL TECHNICAL PUB[illegible] 5

Peter R. Johnston, Ametek, Inc., and
Hans G. Schroeder, International
Consultants Assn., editors

ASTM Publication Code Number (PCN)
04-975002-39

1916 Race Street, Philadelphia, PA 19103

Library of Congress Cataloging-in-Publication Data

Fluid filtration.

(ASTM special technical publication; 975)
"ASTM publication code number (PCN) 04-975001-39."
"ASTM publication code number (PCN) 04-975002-39"—
V. 2, t.p.
Includes bibliographies and indexes.
Contents: v. 1. Gas/Robert R. Raber, editor;
v. 2. Liquid/Peter R. Johnston and Hans G. Schroeder,
editors.
1. Filters and filtration—Congresses. 2. Fluids—
Congresses. I. Raber, Robert R. II. Johnston,
Peter R. III. Schroeder, Hans G. IV. ASTM Committee
F-21 on Filtration. V. Filtration Society (Great
Britain). American Program Committee. VI. Series.
TP156.F5F576 1986 660.2′84245 86-22237
ISBN 0-8031-0926-1 (set)
ISBN 0-8031-0945-8 (v. 1)
ISBN 0-8031-0946-6 (v. 2)

Library of Congress Catalog Card Number: 86-22237

NOTE

The Society is not responsible, as a body,
for the statements and opinions
advanced in this publication.

Printed in Baltimore
October 1986

Foreword

The symposium on Gas and Liquid Filtration was held in Philadelphia 20–22 October 1986. The symposium was sponsored by ASTM Committee F-21 on Filtration and the American Program Committee of the Filtration Society. Peter R. Johnston, Ametek, Inc., and Hans G. Schroeder, International Consultants Association, served as chairmen of the Liquid Filtration sessions, and Robert R. Raber, Farr Co., served as chairman of the Gas Filtration sessions. P. R. Johnston and H. G. Schroeder are coeditors of this publication (Fluid Filtration: Liquid, Vol. II). R. R. Raber is the editor of Fluid Filtration: Gas, Vol. I.

Related ASTM Publications

Fluid Filtration: Gas (Volume I), STP 975 (1986), 04-975001-39

A Note of Appreciation to Reviewers

The quality of the papers that appear in this publication reflects not only the obvious efforts of the authors but also the unheralded, though essential, work of the reviewers. On behalf of ASTM we acknowledge with appreciation their dedication to high professional standards and their sacrifice of time and effort.

ASTM Committee on Publications

Contents

Overview

1986 marks the tenth anniversary of ASTM Committee F-21 on Filtration. Organized on 10-11 March 1976 in Philadelphia, Committee F-21 has worked within the following scope: "The development of test methods, performance specifications, practices, definitions, and classifications, and the stimulation of research to support performance standards for filtration systems, media, and equipment."

In the decade since the committee's founding, significant advances have been made in particle measurement technology and in the overall state of the art in filtration. These facts led the Executive Subcommittee of F-21 to conclude that a symposium on Gas and Liquid Filtration would be most timely. This Special Technical Publication (STP 975) contains 45 refereed technical papers that were reviewed and revised prior to their presentation at the 20-22 October 1986 symposium. STP 975 is organized into Volume I on Gas Filtration with 29 papers and Volume II on Liquid Filtration with 16 papers.

ASTM Committee F-21 on Filtration, in organizing this symposium, provided a forum for those interested in fluid filtration to bring forth comments and questions that might guide the Committee in future efforts. When the Committee was formed in 1976, it immediately broke into two subcommittees, one to address gas filtration and the other to address liquid filtration. This division came about simply because of two separate groups of interest and, perhaps, because of a feeling that laws of nature set the two subjects apart.

Before Subcommittee F21.10 on Liquid Filtration could develop standards, it first had to organize a common language from the jargon of its individual members [Definitions of Terms Relating to Filtration (F 740-82a)]. Some examples of this consensus on terminology are:

1. Since filtration is accomplished by passing a fluid through a porous material, that material is logically called a filter medium. The apparatus containing this medium is called a holder or housing; together the two are called the filter assembly or filter apparatus.

2. The plural of medium is media.

3. Where investigators speak of the flow

characteristics of a filter medium, Subcommittee F21.10 explains that the medium does not flow; instead, fluid flows through it--under a driving force. The ratio of flow to force is referred to as permeability.

4. The use of contaminate for particles is discouraged for two reasons: (1) the particles may be desirable material to be recovered; (2) a contaminate may be a material in solution and thus not filterable. (Adsorption from solution is another subject and is not addressed by F21 even though, in filtration, particles are sometimes attracted to the pore walls of the filter medium by various forces.)

5. Particle is a noun; particulate is an adjective. (Where writers speak of particulates they should more correctly say particulate materials; F21.10 recommends that one should simply say particles and where appropriate, droplets, when an immiscible liquid is present in a carrier liquid.)

6. Porosity does not refer to pore size; it refers to the ratio of void volume within the filter medium to the bulk volume.

This brings us to the meaning of pore size, or the pore-size rating of a filter medium, and to a basic question: What does a writer mean who refers to a filter medium as 0.2 μm or 20 μm? That writer can mean almost anything. To give substance to that meaning, the writer must carefully spell out the method used to arrive at this rating. If the rating method is based on a fluid-intrusion measurement of pore size, the details of that method must be explained. Where pore-size distribution is addressed, the writer must not confuse the median (midpoint) size to the mean (arithmetic average) size. And, where the "largest pore" is addressed, the writer must explain the limit of the distribution which is embraced. If the rating is based on a filtration test, then the details of that test must be explained, such as descriptions of the test fluid and the test particles and, specifically, how particle sizes are measured.

With these considerations in mind, Subcommittee F21.10 wrote three filtration-test standard practices (rather than specific methods), in essence saying to the investigator: After you have decided what test liquid to use and what test particles, proceed as follows ... .

These three practices address: (1) a single-pass, constant-rate test [ASTM Practice for Determining the Performance of a Filter Medium Employing a Single-Pass, Constant-Rate, Liquid Test (F 796-82)]; (2) a single-pass, constant-pressure test [ASTM Practice for Determining the Performance of a Filter Medium Employing a Single-Pass, Constant-Pressure, Liquid Test (F 795-82)]; and (3) a multipass, constant-rate test [ASTM Practice for Determining the Performance of a Filter Medium Employing a Multipass, Constant-Rate, Liquid Test (F 797-82)]. Each of these practices addresses the situation where the investigator may use an automatic particle counter to determine the particle-size distribution in both the feed stream and the filtrate. And, each practice addresses the three alternative ways of expressing how efficiently different-size particles are stopped.

The meaning of particle size is addressed in ASTM Practice for Comparing Particle Size in the use of Alternative Types of Particle Counters (F 660-83).

ASTM Practice F 902 provides a method of calculating the flow-average pore diameter from measurements of porosity and permeability [ASTM Practice for Calculating the Average Circular-Capillary-Equivalent Pore Diameter in Filter Media from Measurement of Porosity and Permeability (F 902-84)].

The subcommittee did write one specific filtration test, F 838, involving the use of *Pseudomonas diminuta* as the test "particles"; but, the method does not suggest the assignment of a rating to the filter medium as a result of test data, nor does it address the question of pass-fail [ASTM Method for Determining Bacterial Retention of Membrane Filters Utilized for liquid Filtration (F 838-83)].

SUMMARY OF TECHNICAL PAPERS

A recent practice to come out of F21.10, and now going to ASTM Society ballot, addresses the separation of liquid droplets from a carrier liquid. From that standards development effort, G. S. Sprenger describes a method of preparing a consistent emulsion. He simply passes a text mixture of oil and water through a globe valve. The other papers in this volume address the separation of particles or microbes from liquid streams.

Johnson offers the only paper on cross-flow filtration, addressing the use of hollow, polypropylene fibers to clarify different beverages and industrial liquids.

Addressing membrane filter media, three separate papers relate the results of integrity tests to separate results of specific filtration tests. That is, Schroeder, Simonetti, and Meltzer show how efficiently different membranes stop two separate test bacteria as a function of integrity tests. Simonetti and Schroeder, in a second paper, test three different membranes as filters for different-diameter latex spheres to offer a correlation between filtration efficiency and separate measurements of pore size. Bower recommends the use of Acholeplasma laidlawii as the test microbe for 0.1-um-rated membranes (adding to the use of other, larger, standard microbes for larger-pore membranes).

Johnston offers theoretical background on the meaning of bubble-point measurements and integrity tests, and why the bubble point (pressure) appears to increase with the thickness of the medium. Wolber and McAllister describe an integrity-testing apparatus.

Levy describes the filtration test performance of a specific membrane in stopping a test bacteria where the carrier water is modified with respect to pH, viscosity, wetting agents, and metal chlorides. That membrane stops that bacteria with great efficiency independent of these variables.

Olson and Greenwood describe the filtration not only of small particles and bacteria but also of fungi and viruses. Filtration efficiency is increased by coating the pore walls with hydrophobic ligands. Lukaszewicz et al. provide a review of the materials of construction of microporous membranes, along with the advantages or use limitations of each.

Johnston discusses general considerations in performing a liquid filtration test. Ostreicher provides the results of test performances of some cartridge filters. Caronia et al. relate the results of mercury-intrusion measurements to results of separate filtration tests of paper media.

Addressing theoretical aspects of porous structures, Johnston offers that, where particles or fibers are packed to build a filter medium, one must question the Kozeny-Carman method of relating changes in porosity to changes in the size of the average pore. Willis et al. propose a general equation addressing the clogging of a filter medium which is applicable to both Newtonian and non-Newtonian fluids. And, Koh offers that the pressure drop of a fluid flowing through a medium can be viewed as either viscous or as inertial losses.

As papers in this volume show, contrasting views of filtration technology persist. This, of course, is not surprising since there are contrasting views even among different ASTM methods. One striking example is the meaning of permeability. One school views permeability as the ratio of fluid flow rate to the driving pressure. The other school views it as the time for a given volume of fluid to flow under a given driving pressure. Thus, where an author would write "filter medium A is more permeable than B", the reader must look to see how that author defines and measures permeability.

Another example of different views is the meaning of pore size in a filter medium. One view sees pore size as that deduced from fluid-intrusion measurements. The other view sees pore size as that deduced from filtration tests. This second view becomes more complicated than the first because of the longer list of variables that must be considered (one obvious variable: the thickness of the medium).

If the users--or potential users--of filter media deduce any "messages" from these papers, we would hope that one message is: Where a filter manufacturer offers a medium with an assigned "micron rating," the user must demand to know how this rating was obtained. And, the user must ask the manufacturer to report permeability, porosity, thickness, and materials of construction.

Peter R. Johnston
Ametek, Inc.
Sheboygan, WI
Symposium cochairman and coeditor

Hans G. Schroeder
International Consultants Assn.
Encinitas, CA
Symposium cochairman and coeditor

Gregory S. Sprenger

ATTEMPTS TO STANDARDIZE FUEL/WATER EMULSIONS FOR THE ASTM COALESCENCE PRACTICE

REFERENCE: Sprenger, G.S. "Attempts to Standardize Fuel/Water Emulsions for the ASTM Coalescence Practice," Fluid Filtration: Liquid, Volume II, ASTM STP 975, P.R. Johnston and H.G. Scroeder Eds., American Society for Testing and Materials, Philadelphia, 1986

ABSTRACT: Drop size and stability of emulsions affect fuel/water coalescer performance greatly. Size and stability of the drops are a function of the liquids, their interfacial nature, and the mixing device used to create the emulsion. Using standard liquids, which are reproducible in physical, chemical, and interfacial properties, drop size is a function of the mixing device solely. The overall objective of this work was to write a coalescer performance test method which addressed the inconsistencies of the mixing device. In this study, Jet A fuel and water were used. Centrifugal pumps, eductors, and valves were tested. The relative size of the drops each produced was measured by a turbidimeter. Centrifugal pumps and eductors were found to create different emulsions as a function of their size. Globe valves were found to create a consistent emulsion, relatively independent of valve size and flowrate. Based on this study, the globe valve throttled to 345 kpad (50 psid) has been specified for the ASTM coalescence practice.

KEYWORDS: emulsion, centrifugal pump, eductors, valves, turbidimeter, mixing device, coalescence

Separation of emulsion liquids has become an important, even critical process. In the marine industry, devices are used to separate the oil from bilge and ballast water for environmental reasons. In the aviation field, it is important to remove water from the fuel before delivery to the aircraft to avoid damaging intricate engine and fuel system components. Failure to remove this contaminant can cause catastrophic results.

Gregory S. Sprenger is a product development engineer at Velcon Filters, 1750 Rogers Ave., San Jose, California 95112.

Over the years, many specifications concerning oil removal from water [1] and water removal from fuels [2,3] have been developed. Many of these specifications exhibited insufficient detail concerning certain important test parameters, such as test emulsion production. For example, MIL-F-8901E [3] allows the use of either a high speed centrifugal pump or flow through a 100 mesh screen. These devices create much different emulsions, with differing droplet sizes; resulting in very different performance characteristics with the same test coalescer/separator system.

The overall objective, which initiated this work, was to write a new ASTM test practice which addressed some of these historical problems, especially related to standardizing the production of the test emulsion.

BACKGROUND

The process of coalescence and separation is used extensively to separate immiscible liquids, such as oil or fuel and water. An emulsion is made up of small droplets dispersed in another immiscible liquid (for example, water droplets in fuel). These small droplets occur naturally in some fuels primarily due to condensation and the mixing action of pumps. This emulsion is pumped through a coalescer [4] to cause drops to collide and grow. They are then usually large enough to settle and separate by gravity due to their density difference.

This process, coalescence, is affected by a number of factors [4, 5, 6]. Davies and Jeffreys [6] believe that coalescence is greatly affected by the relative and absolute sizes of the droplets and by surface active agents, which affect interfacial viscosity and tension; that is, the size and stability of the emulsion. Drop size is critical because smaller droplets require more steps of coalescence to reach a size where gravity settling is effective. Coalescers also capture these smaller droplets less efficiently. These factors make coalescer/separator systems inherently less effective. Emulsion stability affects the actual coalescence rate, causing slower coalescence between droplets, resulting in poorer gravity separation of smaller drops.

These emulsion properties, drop size and stability, are most affected by these factors:

1) The physical properties of the liquids, such as bulk viscosities, densities, and dissolved impurities,
2) The interfacial properties of the liquids, such as interfacial tension, interfacial viscosity, surfactant content, electrolyte content, and
3) The mixing device(s) used to make the emulsion, such as pumps and valves.

In order to produce consistent emulsions, in terms of drop size and stability, and ultimately repeatable coalescence, all these items must be controlled. Items 1) and 2) can be controlled using standard liquids which are repeatable and reproducible in terms of physical and interfacial properties. Such liquids are distilled water, pure organic chemicals, and clay-treated fuels. Item 3) the mixing device, must also be controlled and standardized, which is the objective of this work.

EMULSION MEASUREMENT

The mixing device, to be practical, must not only produce a consistent emulsion, in terms of droplet size and distribution, but be relatively independent of dispersed phase concentration (in this case, water) and flowrate. It must be easily sized for various flow ranges and, hopefully, be relatively inexpensive. In order to evaluate the mixing device, the size and preferably the distribution of the emulsion droplets must be determined. The measurement device must be in-line due to the relative instability of these emulsions.

A 2 year search for a droplet size and distribution measuring instrument proved fruitless. Although these types of devices are used extensively to size solid particles in liquids and aerosols [7], they are not suitable for unstable liquid/liquid emulsions.

An in-line forward scatter turbidimeter (Kaydon Model 861B, Figure 1) was finally tried. In this device, the "cloudiness" of the fuel/water emulsion is detected. The ratio of the scattered light to direct light, as measured by the photo cells, is a measure of turbidity in 'Jackson Turbidity Units (JTU).' These types of devices have been used for water determination previously [8],

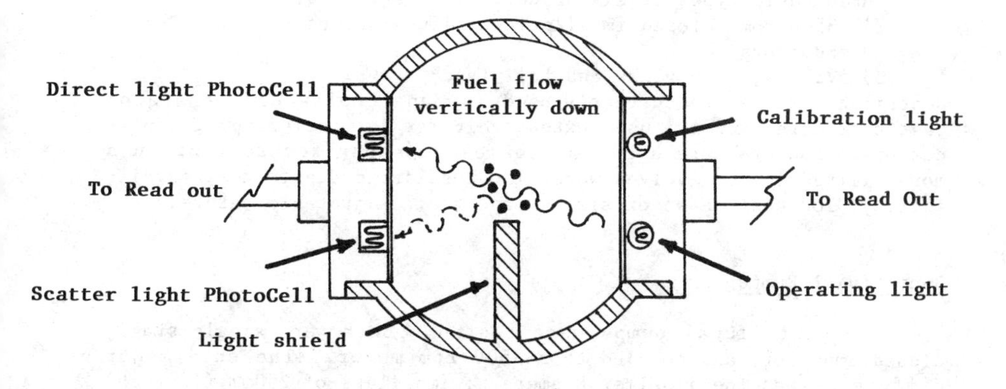

Fig. 1 -- In-line forward scatter turbidimeter

although not in this way. There also exist other ASTM procedures which use turbidimetric devices to determine the particle sizes of solid materials, such as cement [9] and metal powders [10]. These devices measure the "cloudiness" of solutions, which can be related to the concentration and particle sizes of the suspended matter. Although these devices are incapable of providing size distribution information, they can provide an "average" indication of the size of the suspended matter (when concentration is known). Here these readings of turbidity are used to compare the emulsions produced by each mixing device. Higher turbidity (at equal water concentrations) indicates a finer emulsion, consisting of smaller water droplets.

EXPERIMENTAL TESTS

Jet fuel is pumped in a recirculating system through the mixing device. (For the centrifugal pumps, the mixing device itself provides the flow.) Water is injected directly upstream of the mixing device, the emulsion is created, and immediately flows to the turbidimeter, where its turbidity is measured. The water drops are subsequently coalesced and removed. Both water injection rate and flowrate are adjustable. Figure 2 shows a schematic of the test loop.

In general, all turbidity measurements are accurate to ± 15%. This inaccuracy is due to flow fluctuations of both the fuel & water and inaccuracies in the flowmeters, themselves. The turbidimeter itself proved to be very repeatable and reproducible.

Clay-treated Jet 'A' fuel and filtered tap water were used for all testing. Although neither liquid appear to be very "standard" and reproducible, little trouble was encountered in repeating the data.

Three basic types of mixing devices were tested:

1) 3500 rpm, closed impeller centrifugal pumps
2) Eductors
3) Valves, both globe and ball type

Centrifugal pumps are currently used in many coalescer/ separator test procedures and are used extensively for jet fuel pumping applications. Eductors are used for coalescer/separator testing on a more limited scale. Valves were also considered due to their availability, low cost, ease of sizing, and their adjustable nature.

Centrifugal Pumps

Two centrifugal pumps were tested. Both were single-stage, closed impeller, and coupled to a 3500 rpm motor. The only major difference was the impeller diameters; impellers of 250mm (10 inch) and 117mm (4.625 inch) diameter were used.

Tests were conducted to check the affect of pump size, temperature, and repeatability. The fuel flowrates were determined based on the flow curves of each pump. the 250mm impellar pump had a maximum flowrate of 400 gpm and a maximum pressure of 170 psi. The 117mm impeller pump had a maximum flowrate of 50 gpm and a maximum pressure of 45 psi.

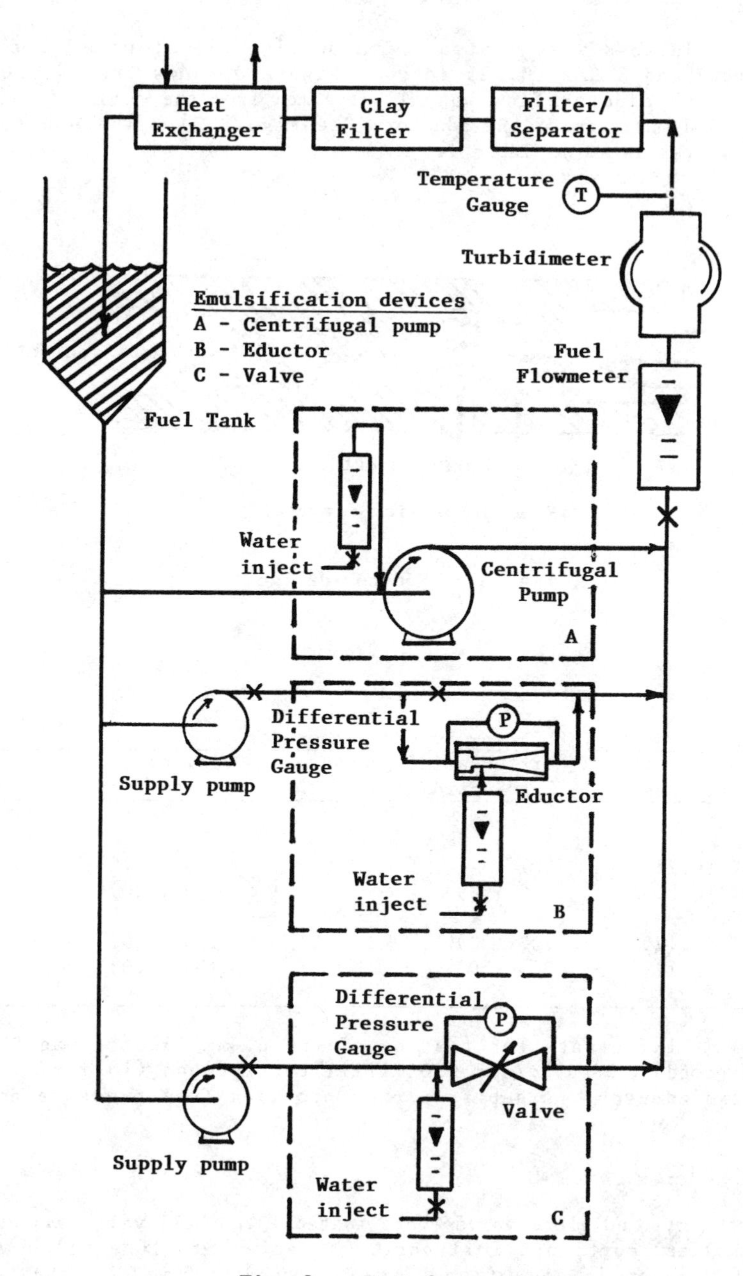

Fig. 2 -- Test loop schematic

Eductors

Two eductors were tested, with orifice diameters of 6.4mm (0.25 inch) and 3.2mm (0.125 inch). Figure 3 shows the eductor design. The eductor flow was slipstreamed from the main flow and recombined downstream of the eductor (See Figure 2). The eductor flowrates are listed in Table 1.

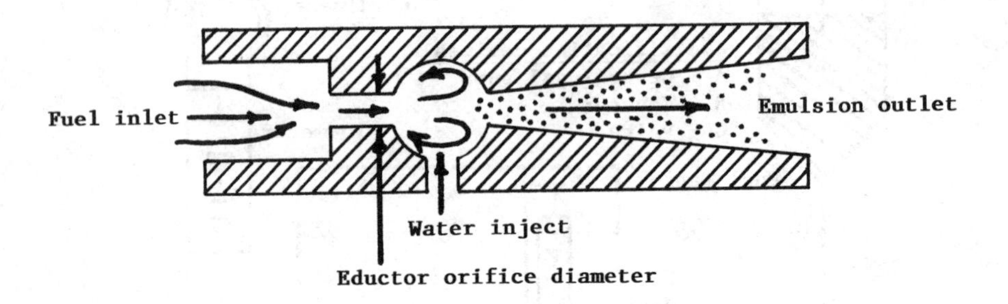

Fig. 3 -- Eductor design

TABLE 1 -- Eductor Flowrates

Eductor size, mm (inches)	Differential pressure kpad (psid)	Flowrate, ml/sec (gpm)
6.4 (0.25)	207 (30)	300 (4.75)
6.4 (0.25)	345 (50)	380 (6.0)
3.2 (0.125)	207 (30)	158 (2.5)
3.2 (0.125)	345 (50)	189 (3.0)

The overall flowrate for testing of the 6.4mm eductor was 2.5 liters/seconds (40 gpm); and 630 milliliters/second (10 gpm) for the 3.2mm eductor. A separate pump provided fuel pressure and flow.

Valves

Both ball and globe valves were tested. The ball valve was standard (smaller) port, not full port. A series of globe vales were tested, ranging in size from 13mm (0.5 inch) to 50mm (2 inch). The globe valves had metal-to-metal seats. Both flowrate through the valve and differential pressure across the valve were varied to test the affect of throttling of the valves. A separate pump provided fuel pressure and flow.

EXPERIMENTAL TEST RESULTS AND CONCLUSIONS

Centrifugal Pumps

Testing was done to check the affect of fuel temperature and pump size, and to determine the repeatability of the test system, including turbidimeter, pump, and liquids. Figure 4 illustrates these results:

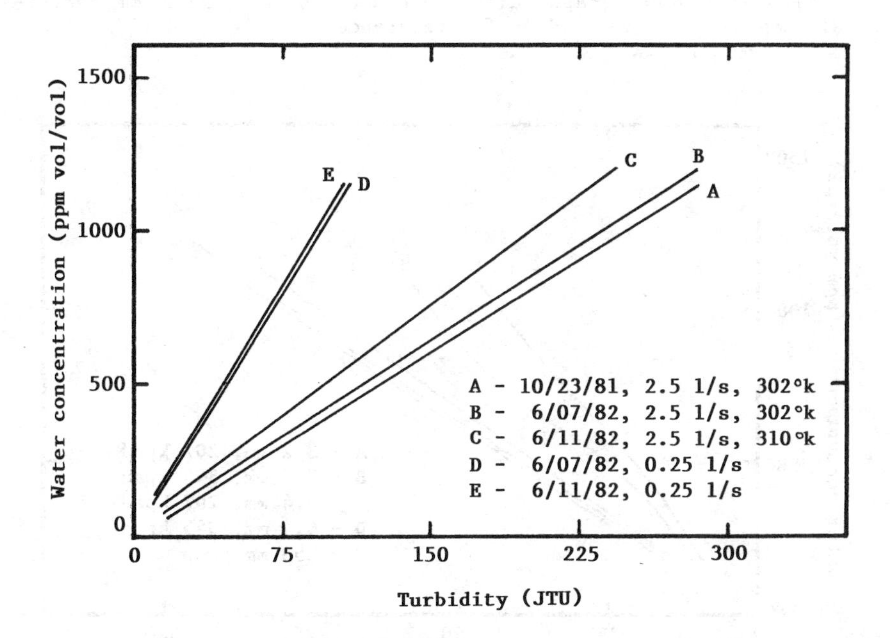

Fig. 4 - Water concentration vs. turbidity, centrifugal pumps

The repeatability of the data is very good. Curve A and B represent data taken almost 8 months apart using the 250mm impeller pump. Repeatability is also very good for the 117mm impeller pump, as shown by curves D and E. This indicates that not only are the pumps repeatable, but so also is the rest of the test system. This would include both fuel and water properties and the turbidimeter.

The temperature of the fuel affects the turbidity/ emulsion size significantly. Curve C indicates lower turbidity and a coarser emulsion. This coarser emulsion may have resulted from coalescence of the unstable emulsion between pump and turbidimeter, or the production of a coarser emulsion in the pump. Additional data, not presented here, indicated little temperature affect between 297°K (75°F) and 308°K (95°F), with greater emulsion changes outside this range. As a result of this factor, the fuel temperature of all subsequent tests were held at 297°K - 300°K (75°F-80°F).

The water concentration vs. turbidity curves are very linear, indicating that the emulsion size is independent of water concentration. That is, as water concentration changes, the droplet sizes remain consistent, only the number of droplets varies. The water concentration vs. turbidity data produced by the two pumps makes them inappropriate as standard mixing devices.

Eductors

Figure 5 shows graphically the eductor data. The 250mm centrifugal pump data has been added for reference.

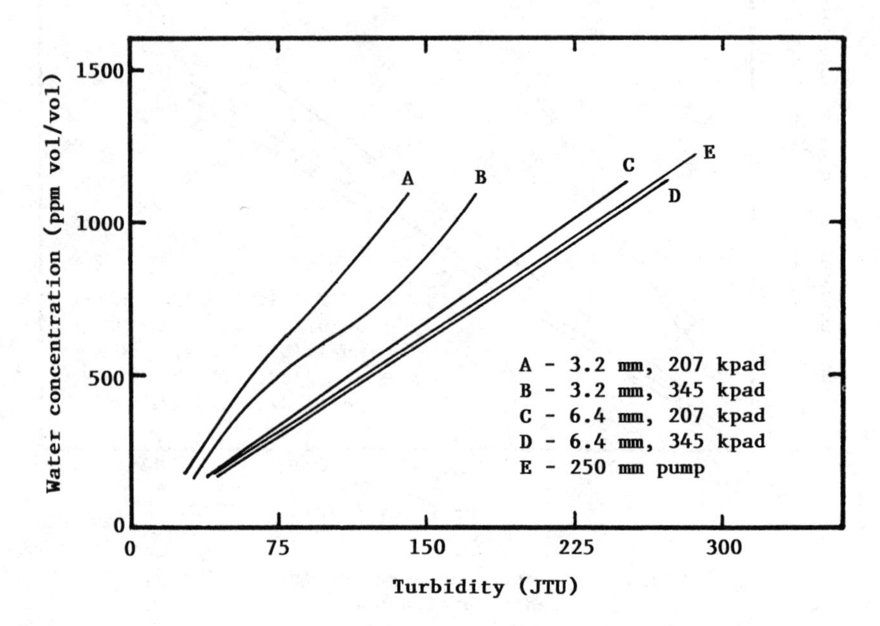

Fig. 5 - Water concentration vs. turbidity, eductors

As was expected, the differential pressure across the eductors (and flow through the eductors) affected emulsion size; the higher differential pressure created a more turbid, finer emulsion. (Data using 69 kpad (10 psid) across the 6.4mm eductor, not presented here, differed even greater.) Comparing curves D and E, the 6.4mm eductor at 345 kpad produced an emulsion similar to that produced by the 250mm centrifugal pump.

Even more important, eductor size greatly affected the emulsion droplet size; the 3.2mm eductor produced a much coarser emulsion at both differential pressures. This is very similar to the pump results.

Again, the wide differences in emulsions produced by the two eductors makes these devices unsuitable. In addition, the requirement of slipstreaming the flow makes them a bit more complicated to operate.

Also noticeable is some curvative in curves A & B. This would indicate that water concentration does slightly effect the drop sizes in the emulsion.

Valves

Initial tests were done to compare the ball and globe valve. Figure 6 shows data gathered with 50mm ball and globe valves at various flows and differential pressure settings. The 250mm pump data has again been added for reference.

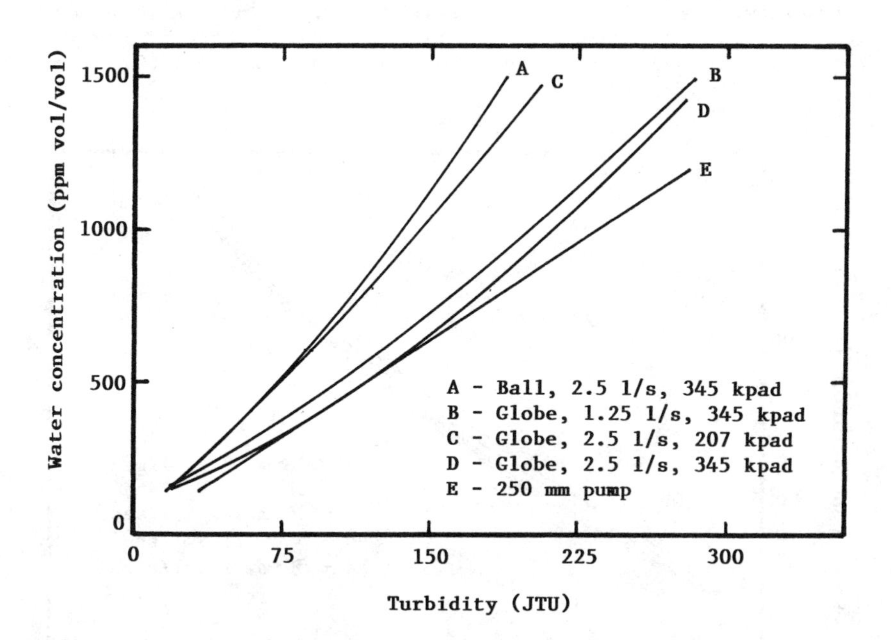

Fig. 6 -- Water concentration vs. turbidity, 50 mm valves

The figure shows, again, that higher valve differential pressures produce finer emulsions. Curves C and D show the same relative differences as data from the eductors.

The difference between the ball and globe valve is evident by comparing curves A and D. The ball valve produces a coarser emulsion at an equal fuel flowrate and valve differential pressure. Comparing the ball valve emulsion data, curve A, with the 250mm pump data, curve E, the ball valve produces a much coarser emulsion. Conversely, curve D, the globe valve data, is close to curve E. The globe valve at 345 kpad (50 psid) produces an emulsion very similar to one produced by the 250mm pump. This indicates that the globe valve is a more efficient emulsion producing device as compared to the ball valve, and produces an emulsion very similar to the desired centrifugal pump emulsion.

The globe valve also appears to be relatively independent of flowrate. Curves B and D differ litle, although curve B represents data gathered at 1.26 liters/sec, half of the flowrate of curve D.

Various sizes of globe valves, ranging in size from 12mm - 50mm (0.5 - 2 inch), were then tested. All of the data presented here was generated with a 345 kpad (50 psid) valve differential pressure. Various flowrates for each valve were checked. Figure 7 shows data for 38mm (1.5 inch) and 50mm (2 inch) valves. Figure 8 shows data for the 25mm (1 inch) valve and Figure 9 shows the 12mm (0.5 inch) valve data.

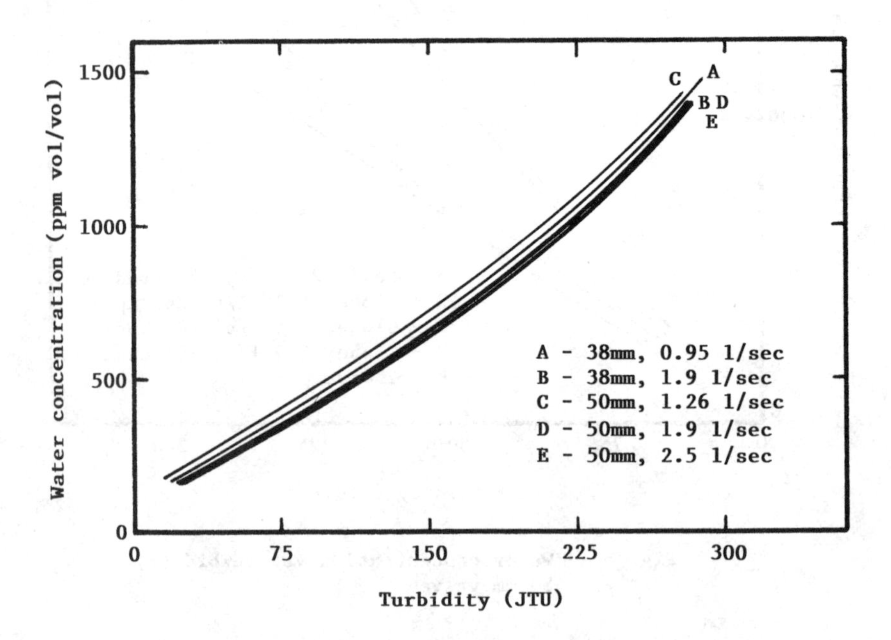

Fig. 7 -- Water concentration vs. turbidity, large globe valves

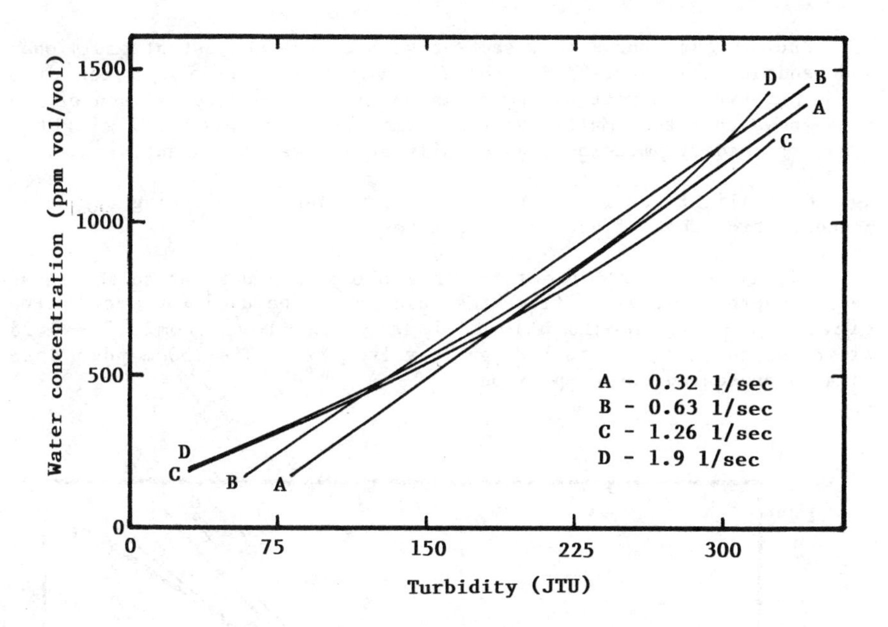

Fig. 8 -- Water concentration vs. turbidity, 25 mm valve

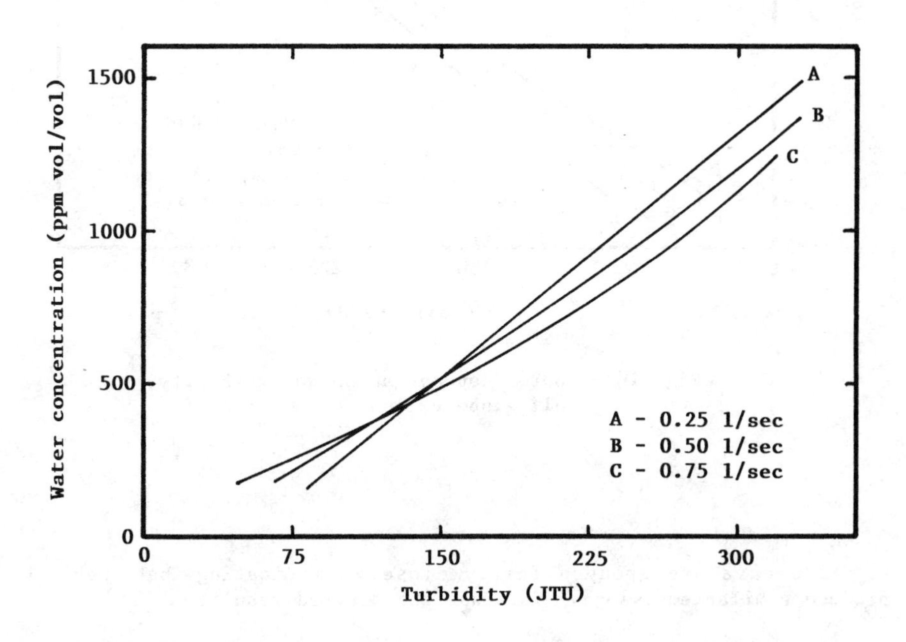

Fig. 9 -- Water concentration vs. turbidity, 12mm valve

Each figure shows that each valve produces consistent emulsions, independent of flowrate. For the 25mm valve, a flow variation of 6 to 1 had little affect on the emulsion size. Close comparison of the data shows that the smaller valves, 12mm and 25mm, produced slightly higher turbidity emulsions, especially at low water concentrations.

Again, a slight curving of the lines is evident, indicating the water concentration does affect the drop sizes.

Figure 10 is a collection of globe valve data at select flowrates, representative of flowrates which might be used for each valve size. This represents a 10 to 1 range in flow, from 2.5 - 0.25 liters/second, and a 4 to 1 range in valve size. The 250mm pump data is also presented for comparison.

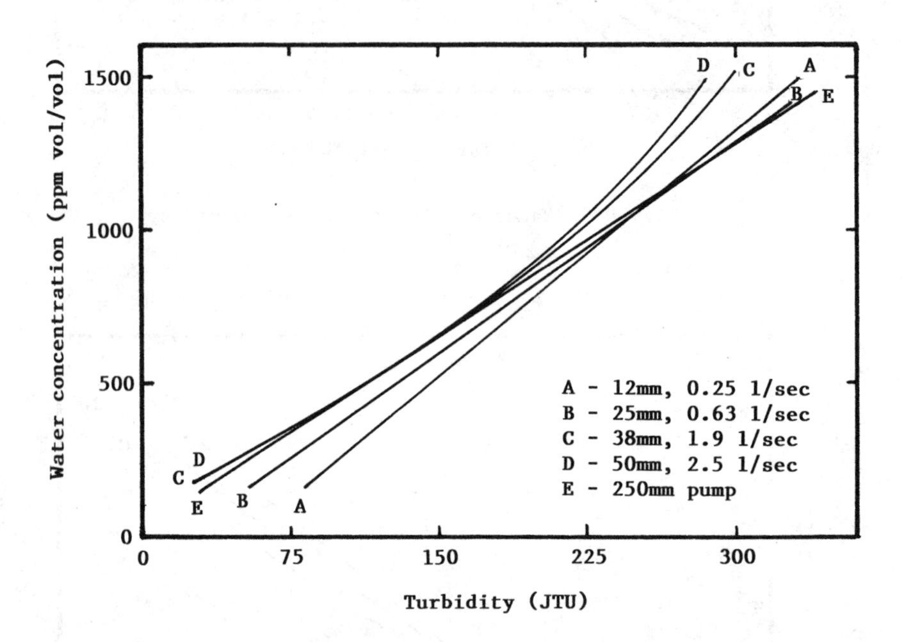

Fig. 10 -- Water concentration vs. turbidity, all globe valves

The data are grouped fairly closely, indicating that they all produce similar emulsions. This was the desired result.

SUMMARY

Neither the centrifugal pumps nor the eductors produced an emulsion with consistently sized drops, with little dependence on flowrate, water concentration, and device size. The ball valve produced an emulsion too coarse in size. The globe valve; however, exhibited the appropriate characteristics. It produced a consistent emulsion, varying little with fuel flowrate through the valve, water concentration injected, and valve size. Drop size (turbidity) was only affected by the differential pressure across the valve; the throttling of the valve. If the valve differential pressure is held to a specified value, a consistent emulsion can be produced to test various sizes of coalescer/separators.

Although in this study, devices were tested in jet fuel and water only, it is believed that other liquid/liquid systems should behave similarly. Although the actual emulsion size may change due to the physical and interfacial properties of the liquids, the globe valve should reproduce a consistent emulsion.

Based on this study, the globe valve, throttled to 345 kpad (50 psid), has been specified for the ASTM coalescence practice.

ACKNOWLEDGEMENTS

The author wishes to thank Kaydon Corporation for the use of the test apparatus and Beverly Waite, Michele Thompson, and Janet Martens for help in preparation of this paper.

REFERENCES

[1] IMCO Specification, "Performance and Test Specifications for Oily - Water Separating Equipment" Spec. A 393 (X) Intergovernmental Maritime Consultative Organization, 1976

[2] "Specifications and Qualification Procedures, Aviation Jet Fuel Filter/Separators" API, Bulletin 1581, American Petroleum Institute, Washington D.C. 2nd. Ed. 1980

[3] "Inspection Requirements and Test Procedures for Filter-Separators, Liquid Fuel: and Filter-Coalescer Elements, Fluid Pressure," Military Specification MIL-F-8901E, U.S. Army 1980

[4] Hazlett, R.N. "Factors in the Coalescence of Water in Fuel" NRL Report 6669, Department of the Navy (Naval Air Systems Command), Washington D.C., 1968

[5] Weatherford, W.D., Jr. "Coalescence of Single Drops at Liquid-Liquid and Liquid-Solid Interfaces,"Report AFAPL-TR-67-3, Air Force Systems Command, Wright-Patterson Air Force Base, Ohio, 1967

[6] Davies, G.A., and Jeffreys, G.V., "Separation of Droplet Dispersions, Part 1 Coalescence of Liquid Droplets," Filtration and Separation Vol. 7, No. 5, September/October 1970, pp. 546-550

[7] Tishkoff, J.M., Ingebo, R.D., and Kennedy, J.B., Eds., Liquid Particle Size Measurement Techniques, ASTM Special Technical Publication 848, American Society for Testing and Materials, Philadelphia, PA, 1984

[8] Mitchell, J., Jr., and Smith D.M., Eds., "Physical Methods," in Aquametry, A Treatise on Methods for the Determination of Water, 2nd Edition, John Wiley and Sons, 1977, pp. 554-612

[9] ASTM Standards, "Test Method for Fineness of Portland Cement by the Turbidimeter," Vol. 04.01 C115 1979

[10] ASTM Standards, "Test Method for Particle Size Distribution of Refractory Metal-Type Powders by Turbidimetry," Vol 02.05 B430 1984

Jeffrey N. Johnson

CROSSFLOW MICROFILTRATION USING POLYPROPYLENE HOLLOW FIBERS

REFERENCE: Johnson, J. N., "Crossflow Microfiltration Using Polypropylene Hollow Fibers, "Fluid Filtration: Liquid, Volume II, ASTM STP 975, P. R. Johnston and H. G. Schroeder, Eds., American Society for Testing and Materials, Philadelphia, 1986.

ABSTRACT: Crossflow microfiltration has become an increasingly attractive process for separation of microparticles, bacteria, emulsion droplets, and for cell concentrating. The efficiency of a crossflow system is dependant upon a number of variables. These include: membrane properties, suspension characteristics, module design, operating parameters, and system design. An optimum combination of these variables can result in an effective separation process.

KEY WORDS: Accurel® polypropylene hollow fibers, periodic backwashing, flux (ℓ/m²/hr), particle boundary layer, crossflow velocity, pore former

INTRODUCTION

Crossflow technology has been successfully used in conjunction with ultrafiltration for many years. However, this technology has spread to microfiltration in recent years.

The alternative to crossflow microfiltration is dead-end microfiltration. This is illustrated in Fig. 1. Dead-end filtration was, until recently, the only alternative available in microfiltration, but dead-end filtration has several drawbacks. As illustrated in Fig. 1, it is handicapped by the rapid decay in flux that occurs due to a continuous build-up of a particle boundary layer on the membrane surface. The particle layer build-up leads to low overall flux rates and requires repeated cleaning or cartridge replacement. These factors result in increased downtime and high operating and cartridge replacement costs.

The crossflow principle (Fig. 2) utilizes shear forces created by a tangential flow across the membrane surface during suspension recirculation to keep particle build-up to a minimum. The particle boundary layer cannot be completely eliminated by crossflow due to

Jeffrey N. Johnson is a chemical engineer working in sales and technical service for Enka America, Inc., Technical Membranes Div., PO Box 1118, Enka, NC 28728.

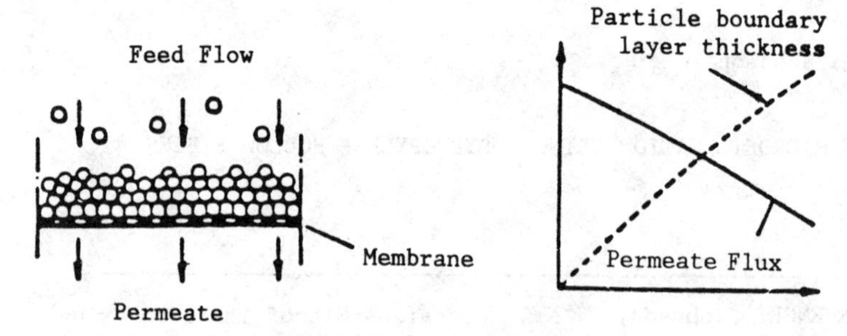

FIG. 1--Dead-end filtration.

the low fluid velocities that exist at the membrane surface. However, by incorporating a suitable membrane into a properly designed module, crossflow filtration can lead to high flux rates with the additional benefit of a reduction in downtime for cleaning.

FACTORS AFFECTING CROSSFLOW MICROFILTRATION

The efficiency of a crossflow microfiltration system is dependant upon a number of factors. The four most significant are:

1. membrane properties
2. module design
3. operating parameters
4. system design

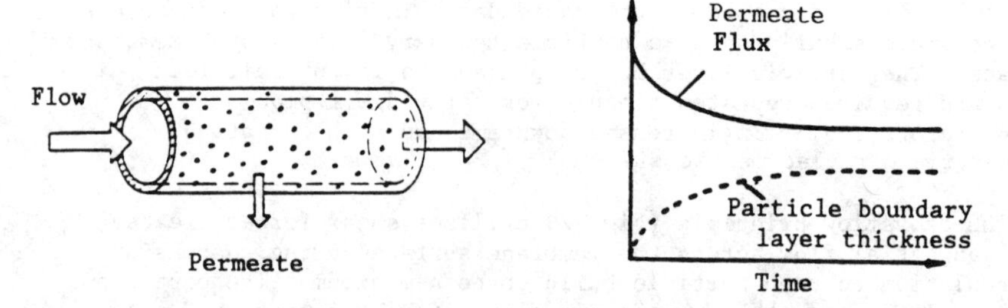

FIG. 2--Crossflow filtration.

Each of the factors must be considered individually and as they influence others in a crossflow system.

Membrane Properties

Various processes are used for manufacturing porous membranes. The polypropylene membranes manufactured by the patented Accurel® process are unique. Although porous membranes can be produced from a variety of thermoplastic polymers using the Accurel® process, polypropylene was chosen due to its strength, chemical resistance, and autoclavability. The unique Accurel® process can be simply described as follows:

> The polymer is dissolved (melted) into a pore forming solution at an elevated temperature to form a homogeneous solution. Upon extrusion of the solution, a regulated reduction in temperature causes the polymer to solidify around the "pore former" thus producing a porous structure. The pore former is then extracted.

The Accurel® process can be used to produce nearly symmetrical membranes with a wide range of pore sizes and porosities with narrow pore size distributions.

The standard Accurel® polypropylene membranes have a nominal pore size of 0.2 μm (based on bubble point testing and bacterial loading tests using pseudomonous diminuta) and a porosity of 70-80%. The high porosity and relatively open surface of these membranes are important because of the high flux rates that result; however, no significant reduction in strength is observed. Table 1 illustrates typical transmembrane pressures at 25°C for some standard Accurel® hollow fibers.

TABLE 1--Typical transmembrane pressures.

	Pressure [Pascal (psi)]	
Inner Diameter (mm)	inside to outside	outside to inside
0.6	2.0×10^5 (29)	1.5×10^5 (21.75)
1.8	1.6×10^5 (23.2)	1.0×10^5 (14.5)
5.5	3.0×10^5 (43.5)	2.0×10^5 (29)

The high porosity of Accurel® polypropylene membranes coupled with the defined flow conditions that exist in hollow fibers and the ability to accommodate large surface areas into a given space make the Accurel® polypropylene membranes ideal for use in crossflow microfiltration. The strength of these membranes offer an additional advantage, periodic backwashing, which will be discussed further.

Module Design

Without proper module design a membrane's effectiveness can be limited. Therefore, compatible housing and potting materials must be chosen to insure resistance against a wide range of chemicals. The modules must be carefully assembled with the hollow fiber membranes securely potted to eliminate leakage. Strict quality control is required to insure module integrity; therefore, Enka's Microdyn® modules are 100% integrity tested.

Crossflow units using hollow fibers can be constructed as complete modules with recirculation and filtrate connections or as cartridges designed to fit into sanitary, stainless steel housings. Microdyn® cartridges are constructed using a polypropylene sleeve and medical grade polyurethane as the potting material and contain either 0.6 mm or 1.8 mm ID hollow fibers and filtration areas of 0.4 to 2.2 m^2. These cartridges can be in-line steam sterilized, therefore they are conveniently suited for critical pharmaceutical and biotechnical applications. The module types contain either 1.8 mm or 5.5 mm ID hollow fibers and filtration areas ranging from 0.036 to 10 m^2. The standard housing material is polypropylene; however, stainless steel housings are available for some applications. All modules containing 1.8 mm ID hollow fibers are potted with polyurethane. Modules containing 5.5 mm ID hollow fibers are potted using polypropylene thus producing a 100% polypropylene module that has excellent chemical resistance and can operate in the pH range from 0.5 to 14.

The cost and lifetime of a crossflow module are key factors in determining the economics of a crossflow system. Frequent module replacement can lead to high operating costs resulting in poor economics for a crossflow system. Accurel® polypropylene membranes have lasted in excess of 2.5 years in continuous operation and have held up through more than 10 autoclavings. As shown in Fig. 3, the cost per square meter of membrane area decreases as module size increases. The availability of large modules to insure economy in industrial size systems is essential since the system cost is proportional to the module cost. General systems cost 2-3 times the module cost. Enka's large 8-10 m^2 modules are ideal for industrial size systems.

Operating Parameters and Flux

The operating parameters set for a specific crossflow filtration system will always depend on the product to be filtered. Optimum parameters must be chosen to provide the highest flux and reliability at minimum cost. The parameters can be determined by mathematical models but these values are usually not suitable for system scale-up. For an accurate scale-up, parameters determined experimentally in small scale pilot plants are necessary. Small modules containing less than 1 m^2 of filtration area are ideal for this purpose. Scale-up, up to a factor of 20 is almost linear.

The most important parameters that are to be optimized include temperature, crossflow velocity, transmembrane pressure, concentration factors and intervals of periodic backwashing. Generally, the higher the temperature the higher the flux achieved. This is usually limited

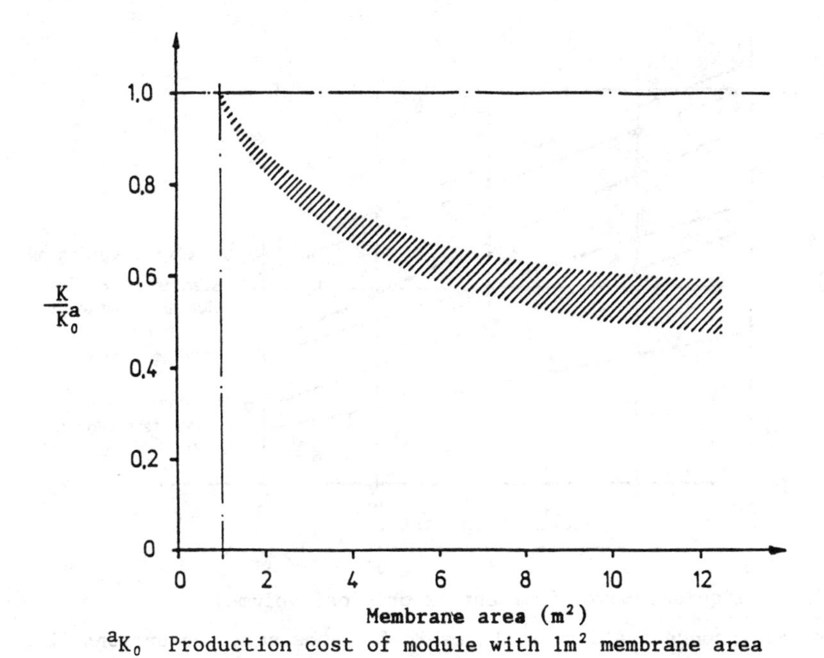

FIG. 3--Decrease in production cost as a function of module size [1].

by the temperature limits of the product being filtered and/or the temperature limitations of the membrane. Increasing the crossflow velocity increases flux rates by increasing shear forces present thus decreasing the particle boundary layer thickness. However, the increase in crossflow velocity results in a higher energy consumption which must be taken into consideration when determining operating costs. The transmembrane pressure has a two-fold effect on the flux. Increasing transmembrane pressure increases flux to a point, but at high pressures particles will tend to embed in the membrane and cause severe decreases in flux and membrane fouling. Usually, the pressures needed to cause fouling are not obtainable due to the pressure limits of the membrane, but this problem must always be considered. Fig. 4 illustrates the effect the concentration factor has on flux. In processes that require high concentration factors, the system can be designed with multiple filtration steps to increase the overall flux and efficiency. Due to their strength, Accurel® PP hollow fibers, are capable of withstanding periodic backwashing. The process entails periodically (at intervals ranging from 2-30 minutes typically) having the filtrate forced back through the membrane in a shock-type manner for 2-3 seconds. Backwashing causes the particle boundary layer to be forced from the membrane surface where it can be swept away by the crossflow current. Fig. 5 illustrates the effect of periodic backwashing on flux during wine filtration. After approximately 70 hours of crossflow filtration, the system was backwashed every 2 minutes for 2 seconds. The effects are obvious.

Fig. 4 shows, in addition to the concentration factor, influence on flux and the different flux rates for a variety of suspensions.

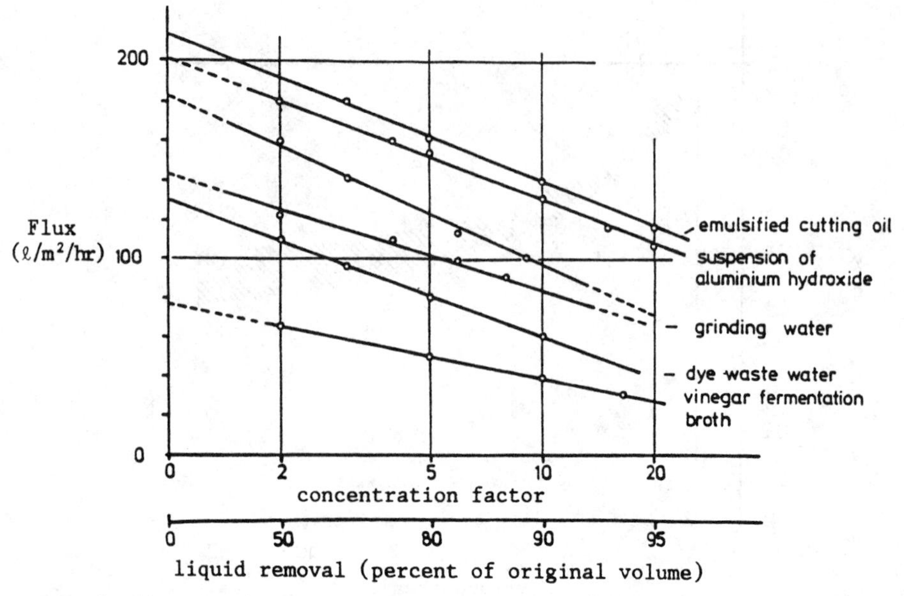

FIG. 4--**Flux-concentration relationship for some real suspensions** [2].

It must be stressed that in some cases impurities e.g., flocculants, surfactants, and proteins can be detrimental to flux rates. This is due to their interaction with the particle boundary layer present.

In summary, there are a wide range of parameters that influence crossflow microfiltration systems. Fig. 6 is a general overview of how these parameters interact directly and indirectly.

Generally, crossflow filtration will be a single step of a process. An illustration of a basic crossflow filtration system utilizing periodic backwashing is shown in Fig. 7. Depending on the application, the design of a crossflow filtration system will vary. However, there are three basic system configurations that can be used. These are:

1. batch

2. continuous

3. multi-stage continuous

The simplest crossflow design is the batch system. Batch systems offer the advantage of achieving the highest overall flux rates obtainable; however, batch systems are limited to smaller applications that can be treated by a discontinuous process. Large scale applications usually require continuous treatment of a suspension. Minor alterations in a batch system are required so that a system can operate continuously. This requires supplying constant product feed at a rate equal to the sum of the flux rate and rate of concentrate removal. The disadvantage of a simple continuous system is that it

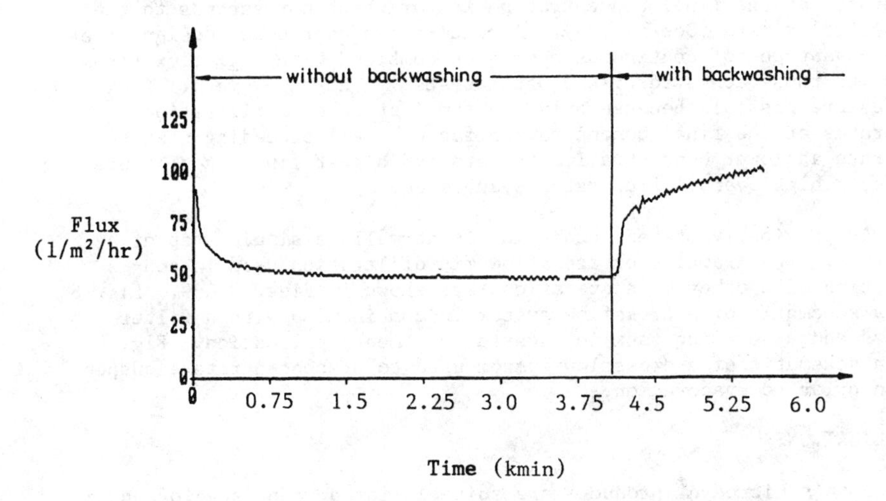

FIG. 5--Wine filtration with and without periodic backwashing [3].

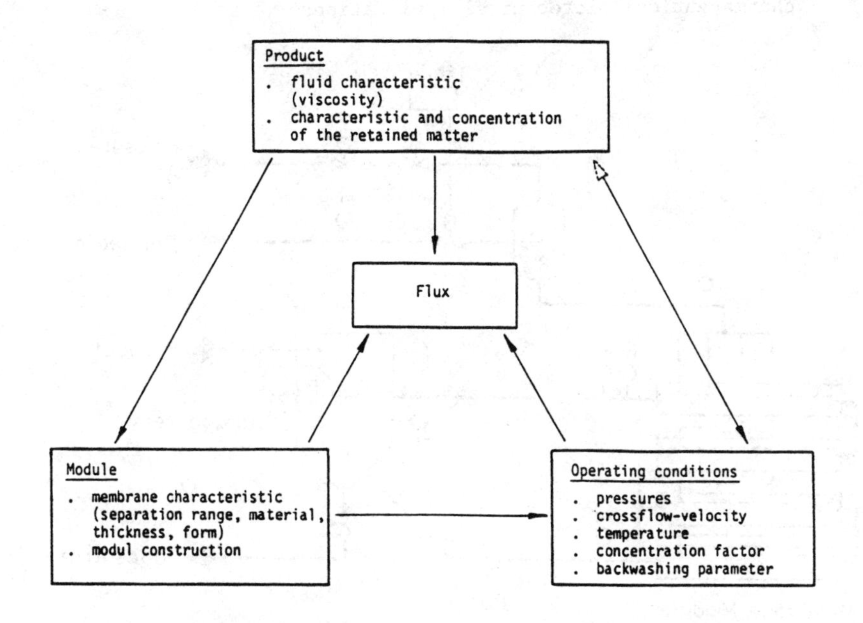

FIG. 6--Important parameters for crossflow microfiltration [4].

operates at the final concentration factor which corresponds to the lowest flux rate (See Fig. 4). A multi-stage continuous design offers the advantages of continuous operation combined with high flux rates near to those achieved with a batch system. The high overall flux rates are possible because only the final stage of a filtration operates at the final concentration factor. All preceding stages operate at lower concentration factors and higher flux rates; therefore, a high overall flux rate is achieved.

As previously stated, crossflow is normally a single step of a process. Two examples of crossflow microfiltration used in conjunction with other unit operations are shown in Figs. 8 & 9. Fig. 8 is a schematic of a crossflow system in combination with a filter press and a settling tank in a waste treatment application. Fig. 9 is a schematic of a crossflow system used to preconcentrate a suspension prior to evaporation.

APPLICATIONS

Enka's Microdyn® products are suitable for a wide range of applications. Generally, these applications can be grouped into one of three categories, which are:

1. food and beverage applications

2. chemical processing applications

3. pharmaceutical/biotechnical applications

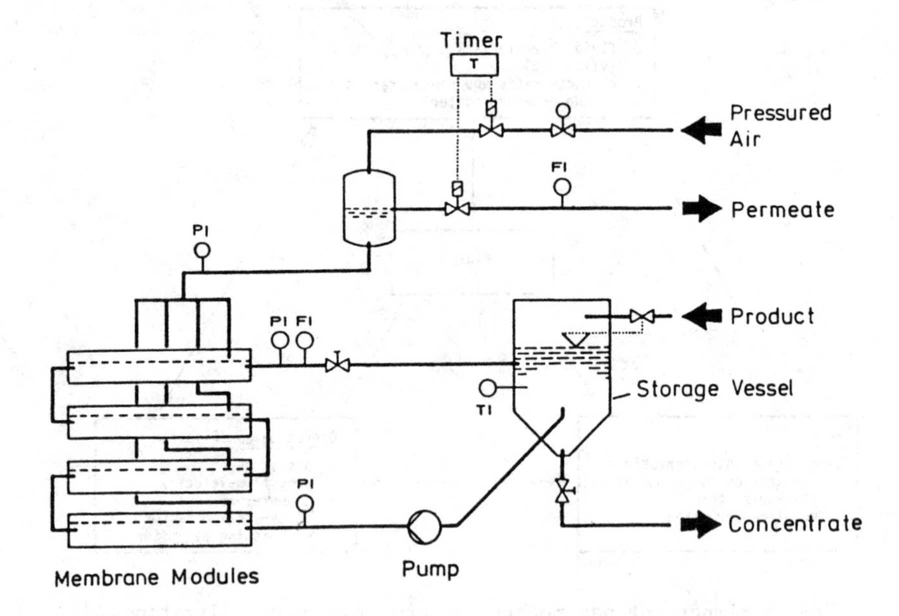

FIG. 7--Schematic drawing of an crossflow microfiltration plant with periodic backwash system.

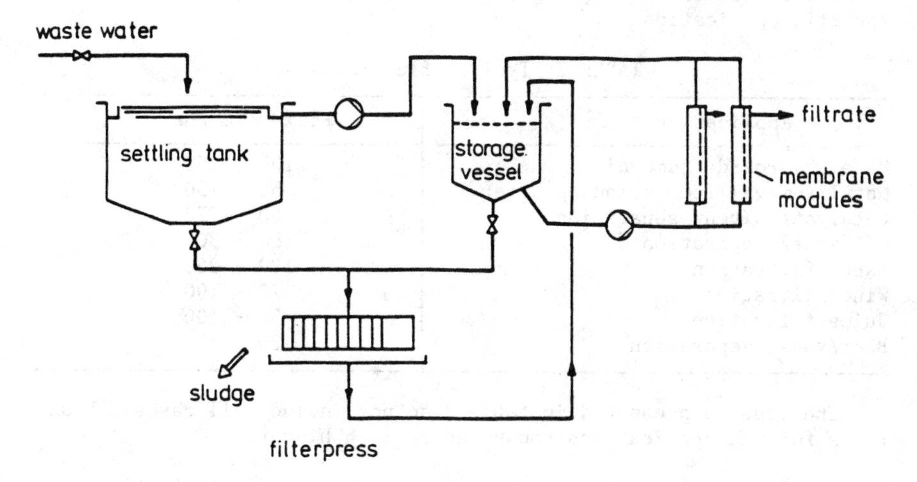

FIG. 8--Crossflow microfiltration as the final cleaning step for waste water.

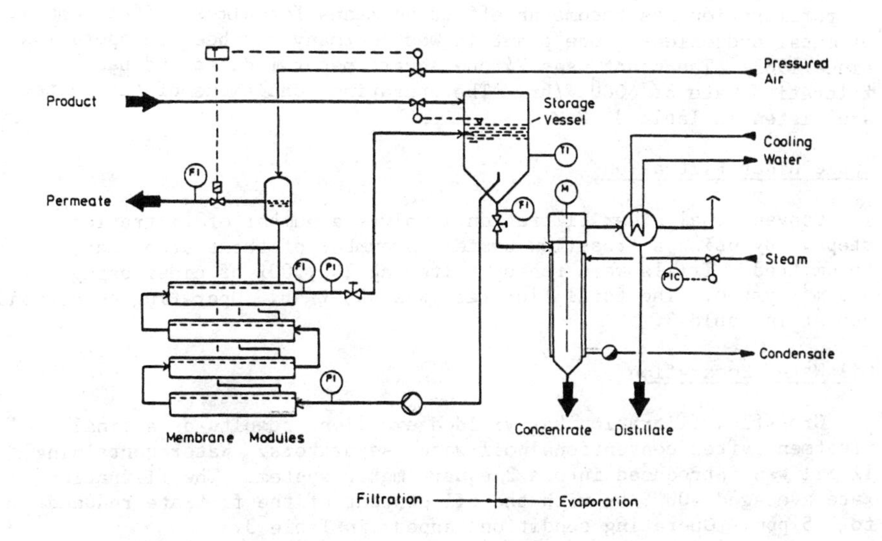

FIG. 9--Crossflow microfiltration as a previous concentration step before evaporation.

Table 2 lists some average flux values obtained for several specific applications.

TABLE 2--Typical flux results.

Application	Flux (ℓ/m²/hr)
Metal hydroxide removal	100 - 250
Metal removal from grinding water	50 - 150
Catalyst/solvent separation	50 - 70
Oil/water separation	150 - 300
Water filtration	100 - 250
Wine filtration	70 - 100
Juice filtration	30 - 100
Beer/yeast separation	20 - 50

The results presented in table 2 do not include all cases. Flux rates for the applications shown can be much higher.

Metal hydroxide filtration, fermented apple cider filtration, and oil/water separation are three applications that lend themselves well to crossflow filtration. Systems used in these areas will be discussed in more detail.

Metal Hydroxide Filtration

With the growing need for more strict waste treatment, crossflow microfiltration has become an effective means for almost total removal of metal hydroxides. One plant in West Germany has been in operation since 1983. The plant uses 24 one square meter modules and has a filtration rate of 6000 ℓ/hr. The operating conditions of the system are listed in Table 3.

Apple Cider Filtration

Conventional cider filtration involves a number of filtration steps. By using a crossflow system, a number of these steps can be omitted. Trials were run by filtering 200,000ℓ of cider using a 8 m^2 system. The total flux rate was 450 ℓ/hr. Operating conditions appear in Table 3.

Oil Water Separation

Crossflow filtration has yielded excellent results as a final treatment after conventional oil/water separators. Water containing 1% oil was introduced into a 2 square meter system. The filtration rate averaged 400 ℓ/hr with the oil content of the filtrate reduced to 3-5 ppm. Operating conditions appear in Table 3.

The applications presented are only a few of the uses for crossflow microfiltration. Additional applications are constantly being discovered.

TABLE 3--Operating Conditions.

Parameters	Application		
	Metal Hydroxide Removal	Apple Cider Filtration	Oil/Water Separation
Operating temp.	20°C	16°C	25°C
Inlet Module pressure	2.0×10^5 Pascal (29 psi)	2.0×10^5 Pascal (29 psi)	1.8×10^5 Pascal (26.1 psi)
Back pulse interval	6 min.	1.5 min.	3 min.
Back pulse duration	2 sec.	1 sec.	2 sec.
Back pulse pressure	2.4×10^5 Pascal (34.8 psi)	2.3×10^5 Pascal (33.4 psi)	2.5×10^5 Pascal (36 psi)
Mean crossflow velocity	2 m/sec.	2.5 m/sec.	3 m/sec.

CONCLUSION

Crossflow microfiltration systems with Microdyn® products can be very effective tools in separation processes with a wide range of applications. The strength and general properties of the Accurel® polypropylene hollow fibers in these products represent a membrane ideally suited for use in combination with crossflow technology. Crossflow microfiltration has found its place as a separation process.

ACKNOWLEDGEMENTS

I would like to express my appreciation to Sigi Oberlaender and Debbie West for their invaluable help in preparing this paper.

REFERENCES

[1] van Gassel, T. J. and Ripperger, S., "Crossflow Microfiltration in the Process Industry," Desalination, Vol. 53, 1985, p. 376.

[2] Ibid., p. 379.

[3] Ibid., p. 381.

[4] Ibid., p. 378.

Hans G. Schroeder, John A. Simonetti and Theodore H. Meltzer

PREDICTION OF FILTRATION EFFICIENCY FROM INTEGRITY TEST DATA

REFERENCE: Schroeder, H.G., Simonetti, J.A., and Meltzer, T.H., "Prediction of Filtration Efficiency from Integrity Test Data," Fluid Filtration: Liquid, Volume II, ASTM STP 975, P.R. Johnston and H.G. Schroeder, Eds., American Society for Testing and Materials, Philadelphia, 1986.

ABSTRACT: The extent of by-pass through potential defects in a filter can be estimated using a simplification of the Standard Method of Test for Pore Size Characteristics of Membrane Filters for Use with Aerospace Fluids (ASTM F-316). A single point gas flow measurement through a wet filter at a test pressure corresponding to the minimum bubble point required for total retention of the contaminant of interest is used to predict the "worse-case scenario". The approach indeed provided a conservative, yet useful prediction of the retention capability of the model filter chosen for the study. While the predicted filtration ratio was about 10^5, observed values ranged from 10^6 to 10^7.

KEYWORDS: filtration efficiency, integrity test, bubble point test, wet flow porosimetry, prefiltration, reduction of bioburden, robust air flow.

INTRODUCTION

Various models such as AC fine test dust, latex or glass beads and carbon fines have been used to describe the retention capability of filters. Although in a real application the efficiency of a filter will depend on the actual contaminant and fluid combination on hand, results generated using such models allow for a general comparison and thus assists in the selection of filter for a given operation.

The ability of a filter to retain a contaminant can be expressed as the ratio of the concentration of the contaminant in the feed stream to the concentration of it in the effluent. The more critical

Hans G. Schroeder is staff scientist for the International Consultants Association, 199 N. El Camino Real #F-318, Encinitas, CA. John A. Simonetti is manager of Laboratory Services for Filtration Technology from Brunswick Technetics, 4116 Sorrento Valley Blvd., San Diego, CA 92121. Theodore Meltzer is senior consultant for Filterite from Brunswick Technetics, 2033 Greenspring Drive, Timonium, Maryland 21093.

the application, the higher this filtration ratio has to be for the operation to be successful. Filtration ratios in the order of 10^{11} (i.e., a retention efficiency of greater than 99.999 999 999%) have been reported and are consistently achieved with filter cartridges used for extremely critical applications such as sterile filtration of injectable pharmaceutical products (1,2,3). This high a level of retention can only be assessed by challenging a filter with a known number (preferably high) of microorganisms (preferably of small size) under predetermined fluid and flow conditions, and subsequently performing a total microbial count on the filtrate recovered.

Since bacterial challenge tests are impractical and destructive in nature, it has become customary to predict the success of filtrative sterilizations based on far more practical physical integrity tests. Correlation of results from bacteria challenge to physical integrity test data is commonly referred to as "validation", a topic which has received wide coverage in literature over the past decade (4,5,6,7). The main objective of integrity testing has been to prevent the use of a defective or the wrong filter for a critical application, but the use of integrity tests as a potential tool for predicting the performance of filters for less critical applications has received little attention.

BYPASS THROUGH A DEFECT

Consider a typical sterilizing grade membrane filter cartridge of 0.5 m^2 filtration area, validated and integrity tested to retain to a close to perfect filtration ratio of higher than 10^{11}. When confronted with a total of 10^9 microorganisms suspended, for the sake of argument, in 100 liters of product to be filtered, chances are very high that indeed the effluent does not contain any of the organisms fed into the filter. Assume that at a reasonable use pressure the flow rate of this "perfect" filter is 10 liters/minute. Experience as well as theory has shown that piercing a hole through the filter medium with a very thin needle will not appreciably improve the flow rate. For argument's sake, let there be an increase of just one milliliter per minute. Such an increase will certainly not affect the time it takes for the filtration to be completed, but by all means one should expect some of the bacteria to pass the filter. As long as the filter does not plug to an appreciable extent (a realistic assumption for small, relatively clean batch operations), the amount of bypass through the pinhole can be estimated to be about one milliliter for every 10 liters, i.e., a total of 10 milliliters for the entire batch. If the nature of the artificially induced defect is such that none of the bacteria reaching it will be retained, the filtrate will contain the 100,000 organisms contained in the 10 ml that remained unfiltered, thus as a consequence of the defect, the filtration ratio dropped from $>10^9$ to $10^9 \div 10^5$ or 10^4. While this 10,000-fold reduction in concentration is far too unreliable for sterilizing injectable products, it might be more than sufficient to take the haze caused by a high concentration of submicrometer particulate matter out of vegetable oil or shampoo to make that type of product aesthetically more appealing to the consumer. A large rip in addition to the pinhole in the filter might render it useless even for the latter application, thus, some type of integrity test for any application may be advisable.

PHYSICAL INTEGRITY TEST

Detailed presentations of applicable non-destructive tests to predict the retention characteristics of membrane filters have been described in literature (8,9,10).

The theoretical basis for physical integrity testing of microporous filter media lies in the fact that wet filter structures are impermeable to bulk flow of a test gas until a pressure high enough to dislodge the liquid from the pores is reached. By describing the filter as a manifold of capillaries, it can be calculated that the radius of the pore blown open is inversely proportional to the test pressure applied, thus the bubble point, as this pressure is referred to is indicative of the largest pores present in the filter tested. At pressures below the bubble point, the test gas will cross the filter at a far slower rate, governed by diffusion. Its magnitude, in accordance with Ficks law will be directly proportional to the test pressure applied.

As long as filters are of equal area, wet with the same liquid, tested with the same gas, have the same thickness and void volume (i.e., porosity, not pore size) their diffusive properties will be identical and variation from this common diffusive flow will depend on the presence of pores open at a given test pressure, as described earlier. As the test pressure is increased, smaller pores are being opened up to the total flow, which for most common filters increases rapidly at pressures over the bubble point. Quantitative monitoring of the gas flow is the basis for more sophisticated integrity tests, since the transition from diffusive to bulk flow provides a wealth of information about the pore size distribution as well as the integrity of the filter (11, 12).

The discussion can be summarized in the profiles shown in Figure 1, which schematically represent the flow of test gas through a wet structures as a function of the test pressure applied. The onset of the curve represents the diffusive flow, which follows the same slope for filters of equal geometry, material, porosity and thickness. As pressure A is reached, the coarsest of the filters starts to show signs of bulk flow of test gas as indicated by the sharp increase in the slope of the profile. For the tighter filters of the same family, this occurs at pressures corresponding to the intrinsic bubble point of the medium, a value which ideally will be indicative of a size above which contaminants will retain consistently.

It stands to reason that the media described by the profiles in Figure 1 will exhibit some rational retention pattern. While medium A may not be able to retain a given size contaminant to an appreciable extent, medium B should be able to perform better, but it may take a medium as fine as C to retain the contaminant totally, thus, making even tighter medium D unnecessary or an overkill for the retention of the contaminant. If total retention is crucial to the operation, obviously, a medium as tight or tighter than C should be chosen and integrity test criteria be set accordingly (13).

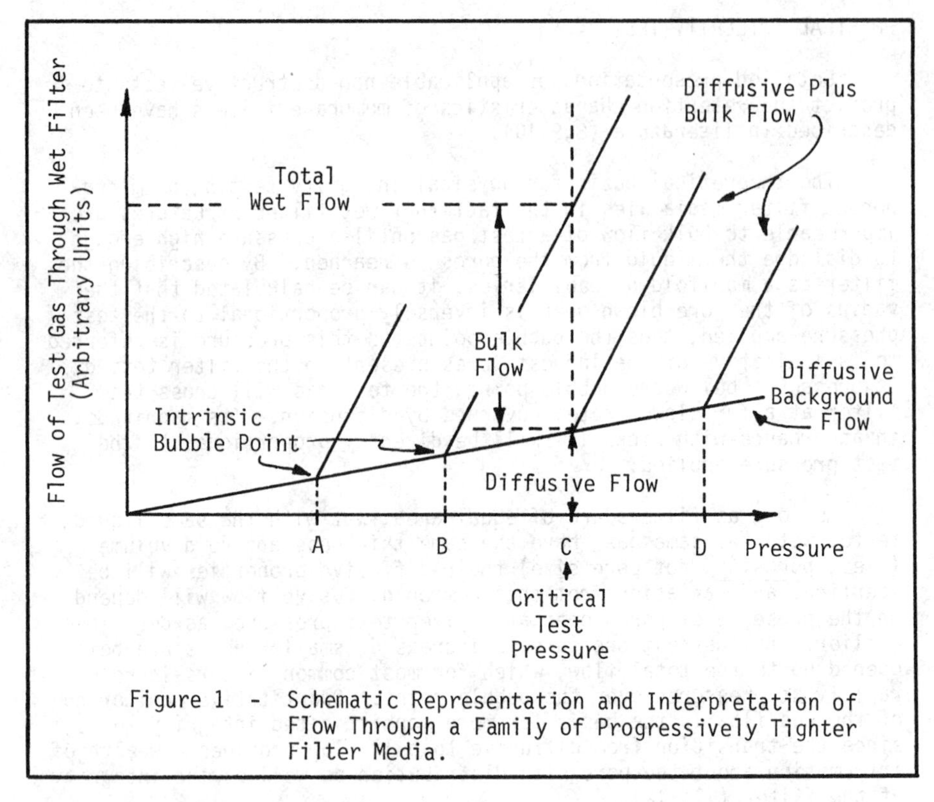

Figure 1 - Schematic Representation and Interpretation of Flow Through a Family of Progressively Tighter Filter Media.

The extent to which coarser media of the same family would be able to retain the contaminant can be estimated from a simple integrity test performed at the critical bubble point pressure, i.e., the pressure C in the above discussions. As can be seen from Figure 1, the total flow of test gas through the wet element consists of a bulk flow added to the diffusive background flow. The lower the bulk flow, the fewer the number of pores open or dry at the critical pressure, thus of the critical size. The fraction of the filters total flow capacity over which the pores are larger than this critical size can be estimated by dividing the value of the bulk flow by the total potential test gas flow that the filter could carry at the test pressure. This ratio corresponds roughly to the by-pass ratio discussed in the pinhole example cited earlier, hence corresponds to the fraction unfiltered. The filtration ratio as defined before is simply the inverse of this unfiltered fraction. Obviously, the gas flow values necessary for the calculation need to be established.

The total gas flow through the wet filter can easily be determined experimentally by direct measurement using an appropriate flow meter, or simply collecting the gas in an inverted graduated cylinder sealed by means of a water trough over a known time period. From this value, the diffusive flow has to be subtracted to determine the bulk flow of gas.

There are three methods for estimating the diffusive flow of a filter. A value can be measured at a pressure below the true bubble point of the filter and adjusted by linear extrapolation to the pressure established for the integrity test. Alternatively, an approximation of the diffusive flow can be calculated from the test pressure, the properties of the gas and liquid involved in the test, and the area, porosity and thickness of the filter (8,9). Last, the contribution of the diffusive flow can simply be ignored since it is small compared to the bulk flow. In the pinhole example cited, a typical diffusive flow value at the test pressure would be 10 ml/min, while the flow through the pinhole is about 1000 ml/min.

Since the bulk flow of the test gas through the wet filter is likely low enough not to be affected by the construction of the filter itself or restrictions due to the housing, the flow rate obtained experimentally can be used for the calculation. However, the total gas flow through the dry filter needed for the calculation of the potential by-pass ratio should not be measured at the test pressure, but rather extrapolated from very low pressure flow readings, since it is the potential (rather than actual) gas flow that characterizes the filter's carrying capacity. For example, the potential total gas flow through the dry filter is in the order of 10^7 standard ml/min, thus the filtration ratio would be 10^7 ml total ÷ (1000 bulk - 10 ml diffusion) which as before amounts to 10^4. In spite of the fact that the example is somewhat arbitrary, the values cited are reasonably realistic.

BYPASS AND RETENTION CHARACTERISTICS OF OVERSIZED PORES

The single point integrity test described obviously yields little, if any, information about the size or the number of the defects that may be present, but this does not limit the predictive capabilities to a great extent. If we consider the filter to be an array of capillaries connected by a manifold, the flow rate through each capillary as a function of its diameter, d, can be estimated using the well established Hagen - Poiseuille equation for laminar flow

$$Q = \frac{\pi \; \Delta P g}{128 \; \eta L} \; d^4$$

where respectively, ΔP, g, η and L are the differential pressure across the capillary, the acceleration of gravity, the viscosity of the fluid and the length of the capillary (approximately the thickness of the medium). Since due to manifolding these values are constant, the equation can be reduced to

$$Q = K \, d^4$$

This expression states that the flow through 256 flaws of diameter d is equal to the flow through 16 holes of diameter 2d, or a single defect of diameter 4d, since

$$K \, (4d)^4 = 16 \, K \, (2d)^4 = 256 \, K \, (d)^4, \text{ etc;}$$

thus the filtration ratio for either of these would be the same providing d is small enough to allow passage of the contaminant in question.

Although a large single defect in a filter medium is possible, a distribution of flaws is more likely to be responsible for passage. This could well be the "tail" of the pore size distribution inherent to the filter, and thus include a large number of pores which are not too much larger than the critical size required for retention. If that is indeed the case, the outcome of the filtration will likely be better than predicted from the integrity test. This can be anticipated from the retention patterns observed in actual filtration situations.

One of the more evident modes of retention occurs when a given particle is physically too large to pass through the specific pore it encountered during its attempted passage through the filter. It is this type of sieve mechanism which can be predicted by flow channelling considerations and integrity test procedures described. While certainly a large enough pore is a prerequisite for passage of rigid particles through a filter, it has long been recognized that a particle small enough to penetrate the medium does not necessarily emerge in the effluent. In addition to mechanical or geometric restrictions, non-mechanical interactions between the particle and the pore can lead to capture of particles despite their small size. The forces which retain the particle are generally due to charge distribution effects, such as hydrogen bonding or other dipole-dipole interactions. It has to be recognized that such modes of capture depend on the nature and properties of fluid filtered, as well as the flow conditions imposed during filtration. In addition, these mechanisms are reversible and subject to saturation, but in spite of this, they will assist the mechanical retention, thus, the efficiency as calculated from the integrity test may be considered somewhat pessimistic.

Two more reasons must be cited as causes for the real retention efficiency to likely be higher than predicted by physical tests. The first is particularly true for asymmetric membranes recently described in literature (14, 15). As the test pressure is increased, the thickness of the liquid layer within the membrane becomes thinner, thus, the diffusive flow of an integral filter will be faster than the linear dependency on pressure described for isotropic media.

The remaining reason for optimism was brilliantly postulated in the "Williams Model" (16, 17). In an integrity test situation, the gas is able to "leisurely" find the largest outlet pore of the individual polyhedral open walled cells that make up the reticulated structures of membrane filters produced by conventional casting techniques. These pores are part of a "maze path" that can turn in any direction, even back toward the pressurized upstream of the membrane. When in use, however, the liquid filtered is physically driven along the shorter, far more direct "tortuous path" across the membrane structure.

RESULTS

In order to assess the applicability of the proposed method, a series of 0.5 m^2 single layer membrane prefilter cartridges were integrity tested and subsequently challenged using recognized procedures. Serratia marcescens was used as a model contaminant for cartridges with a 0.45 micrometer nominal retention rating, while cartridges rated at nominal 0.2 and 0.1 micrometers were challenged with Pseudomonas diminuta. For the filter media under investigation, i.e., highly asymmetric polysulfone membranes, the critical bubble points for "total" retention of these organisms is 22.5 and 30 psig (1 psid = 6.76 kPa) respectively, thus the integrity test was performed at these pressures. Table 1 summarizes the results obtained. The observed filtration ratio was calculated by dividing the known total challenge level by the number of organisms recovered in the effluent. The value listed for the total dry air flow was extrapolated from low pressure "typical" consistent values, and divided by the total measured wet flow to arrive at the predicted filtration ratio. No correction was made to establish the true bulk portion of the measured wet flow. The accuracy of the prediction is expressed as the ratio of the actually obtained to the predicted retention listed in the last column of Table 1.

DISCUSSION

Considering the relatively high wet flow values observed, it can be assumed that a bubble point well below the present test pressure should be observed for these cartridges. Since the object of such a bubble point test is assuming total retention, rejection of these filters would indeed have been warranted in view of the passage observed. Obviously, the passage does not come as a surprise, since the filters tested are not designed to be of sterilizing grade or intended for use in such extremely critical applications. With this in mind, the retention efficiencies in the order of better than 99.9999% can be considered remarkable rather than failures in the performance of these filters.

The results of the experiments clearly indicate that the actual retention performance is superior to what would be predicted by the proposed integrity test by a factor of about 10 to 100. As discussed earlier, this is in part due to the asymmetry of the membrane, since the thinning out of the water layer causes the diffusive flow to become a larger than expected portion of the wet flow observed. No doubt that for conventional isotropic media, the prediction would be more accurate. It is also conceivable that at a higher flow rate or suspending the microorganism in another vehicle could have resulted in passage of more organisms, which would have caused the prediction to be closer than the value obtained in these experiments. In addition to this, it has to be recognized that the filter is likely to plug faster than the defects, thus, a portion of the initial safety margin will be lost during use of the filter. In view of these considerations, the ability to predict the retention to within one or two orders of magnitude may well be realistic and useful.

In principle, a similar test under the name of "robust air flow" was proposed several years ago as a replacement for the somewhat subjective bubble point test performed in conjunction with filtrative sterilization (18). From the data presented here, it seems unlikely that such a test is sensitive enough to predict total retention. In fact, inadequate test sensitivity could have been at play in the

Table I: Comparison of Predicted and Experimentally Obtained Filtration Ratios.

Nominal Rating (um)	Test Pressure (psig)	Total Dry Air Flow (ml/min)	Total Wet Flow (ml/min)	Filtration Ratio Predicted	Total Challenge Level	Number of Organisms in Effluent	Filtration Ratio Observed	Accuracy of Prediction
0.4[a]	22.5	$5.0x10^7$	394	$1.27x10^5$	$7.4x10^9$	145	$5.1x10^7$	40
			96	$5.20x10^5$	$8.0x10^9$	250	$3.2x10^7$	61
			384	$1.30x10^5$	$9.3x10^9$	118	$7.9x10^7$	61
			153	$3.26x10^5$	$9.4x10^9$	448	$2.1x10^7$	64
			132	$3.79x10^5$	$8.0x10^9$	198	$4.0x10^7$	105
			206	$2.43x10^5$	$8.0x10^9$	290	$2.8x10^7$	115
0.2[b]	30.0	$4.5x10^7$	141	$3.19x10^5$	$1.9x10^8$	74	$2.6x10^6$	8
			86	$5.23x10^5$	$2.4x10^8$	18	$1.3x10^7$	19
			34	$1.30x10^6$	$3.0x10^8$	10	$3.0x10^7$	23
			94	$4.79x10^5$	$2.3x10^8$	11	$2.1x10^7$	44
0.1[b]	30.0	$2.8x10^7$	115	$2.44x10^5$	$2.4x10^8$	110	$2.2x10^6$	9
			195	$1.44x10^5$	$2.4x10^8$	115	$1.1x10^6$	15
			145	$1.93x10^5$	$2.4x10^8$	33	$7.3x10^6$	38
			33	$8.48x10^5$	$2.9x10^8$	6	$4.8x10^7$	57
			94	$2.98x10^5$	$2.3x10^8$	11	$2.1x10^7$	70

[a] Serratia marcescens per 0.5 m^2 cartridge

[b] Pseudomonas diminuta per 0.5 m^2 cartridge

reported cases of bacteria passage through supposedly integral filters (19, 20). However, the proposed test may very well be suited for the many applications for which better than nominal retention performance is crucial to the success of a given process.

REFERENCES

(1) "Microbiological Evaluation of Filters for Sterilizing Liquids," Health Industry Manufacturers Association, Washington DC, HIMA Document No. 1, Vol. 4 (1982).

(2) Wallhausser, K.H., "Bacterial Removal Filtration with Highly Asymmetric Filter Media." Pharma International, Dec. 82, Feb 83 and Apr. 83.

(3) Johnston, P.R. and Meltzer, T.H., "Comments on Organism-Challenge Levels in Sterilizing Filter Efficiency Testing," Pharmaceutical Technology, Vol. 3, No. 11, Nov. 1979, p. 66.

(4) Leahy, T.J., and Sullivan, M.J., "Validation of Bacterial - Retention Capabilities of Membrane Filters," Pharmaceutical Technology, Vol. 2, Nov. 1978, p. 65.

(5) Pall, D.B., and Kirnbauer, E.A., "Bacterial Removal Prediction in Membrane Filters," 52nd Colloid and Surface Symposium, University of Tennessee, Knoxville, TN, June 12, 1978.

(6) Olson, W.P., Martinez, E.D., and Kern, C.R., "Diffusion and Bubble Point Testing of Microporous Cartridge Filters: Preliminary Results at Production Facilities," Journal of Parenteral Science and Technology, Vol. 35, No. 5, 1981, p. 215.

(7) Wallhausser, K.H., "Validierungs-Verfahren zur Uberwachung der Entkeimungs-Filtration," Pharm. Ind., Vol. 44, No. 4, 1982, p. 401.

(8) Reti, A.R., "An Assessment of Test Criteria in Evaluating the Performance and Integrity of Sterilizing Filters," Bulletin Parenteral Drug Association, Vol. 31, 1977, p. 187.

(9) Schroeder, H.G., and Deluca, P.O., "Theoretical Aspects of Sterile Filtration and Integrity Testing," Pharmaceutical Technology, Vol. 4, No. 11, 1981, p. 80.

(10) Hofmann, F., "Integrity Testing of Microfiltration Membranes," Journal of Parenteral Science and Technology, Vol. 38, 1984, p.148.

(11) Schroeder, H.G., and Deluca, P.O., "A Rational Approach to Integrity Testing of Filter Cartridges," Interphex 80, Section Research and Development, Sept. 15-17, 1980, p. 62.

(12) Johnston, P.R. Lukaszewicz, R.C., and Meltzer, T.H., "Certain Imprecisions in the Bubble Point Measurement," Journal of Parenteral Science and Technology, Volume 35, No. 1, 1981, p. 36.

(13) Schroeder, H.G. and Simonetti, J.A., "Selection of Integrity Test Parameters for Sterilizing Grade Filters," submitted to Journal of Parenteral Science and Technology.

(14) Kestings R, Murray, A, and Newman, J., "Highly Anisotropic Microfiltration Membranes," Pharmaceutical Technology, Vol. 5, May, 1981, p. 53.

(15) Wrasidlo, W. and Mysels, K.J., "The Structure and Some Properties of Graded Highly Asymmetrical Porous Membranes." Journal of Parenteral Science and Technology, Vol. 38, No. 1, 1984, p. 25.

(16) Williams, R.E. and Meltzer, T.H., "Membrane Structure, the Bubble Point and Particle Retention: A New Theory," Pharmaceutical Technology, Vol. 7, May 1983, p. 36.

(17) Williams, R.G., "Absolutely Accidental Absolute Filtration," SME Lecture, Filtration Course, 1985.

(18) Johnston, P.R. and Meltzer, T.H., "Suggested Integrity Testing of Membranes Filters at a Robust Flow of Air," Pharmaceutical Technology, Vol. 4, Nov, 1980, p. 49.

(19) Howard, G., and Duberstein, R., "A Case of Penetration of 0.2 um Rated Membrane Filters by Bacteria," PDA Paper November 1979 meeting.

(20) Wallhauser, K.H., "Is the Removal of Microorganisms by Filtration Really a Sterilization Method?", Journal of Parenteral Drug Association, Vol. 33, No. 3, 1979, p. 156.

John A. Simonetti and Hans G. Schroeder

PARTICLE RETENTION OF SUBMICROMETER MEMBRANES

REFERENCE: Simonetti, J.A., Schroeder, H.G., "Particle Retention of Submicrometer Membranes, "Fluid Filtration: Liquids, Volume II, ASTM STP 975, P.R. Johnston and H.G. Schroeder, Eds., American Society for Testing and Materials, Philadelphia, 1986.

ABSTRACT: Various methods used to determine the pore size of filter media were applied to several sterilizing grade microporous membranes. In spite of the fact that the microbial retention rating for the samples evaluated is 0.2 micrometers, differences were observed in the calculated viscous-average-flow pore diameter, the maximum pore as determined by the bubble point and the retention characteristics of known size particles. Although the biological approach to rating filters has served the pharmaceutical industry rather well over the past decades, it is evident that the rating can be misleading for industries that count on small particle retention. To illustrate this point, the latex retention characteristics of 0.2 micrometer rated filter cartridges are included in the evaluation.

KEYWORDS: Retention rating, micrometer rating, latex particle retention, wet flow porosimetry, viscous-average-flow pore, pore size analysis.

INTRODUCTION

The main purpose for most filtration operations is the removal of undesirable contaminants from process streams. The more critical the application, the finer the filter rating is generally chosen to achieve the desired result. Since many of the submicrometer applications intend to remove microbes from solutions, a consensus has been reached to rate the filter's removal capability against model microorganisms. *Serratia marcescens* is used as per ASTM Designation: 3863-80 to verify a 0.45 micrometer rating while the smaller *Pseudomonas diminuta* has been used to verify the more demanding 0.2 micrometer requirements as per ASTM Designation:3862-80. This

John A. Simonetti is manager of Laboratory Services for Filtration Technology from Brunswick Technetics, 4116 Sorrento Valley Blvd., San Diego, CA 92121. Hans G. Schroeder is staff scientist for the International Consultants Association, 199 North El Camino Real #F-318, Encinitas, CA. 92024.

approach to rating is associated with the benefit of a very high sensitivity, and certainly routinely achieved titre reductions of 10^{11} testify to the degree of perfection of these filters when confronted with bacteria in short term challenges. Although this approach to rating has served the pharmaceutical industry rather well over the years, performance discrepancies in longer term challenge results in grow-through of bacteria through some of the 0.2 micrometer rated membranes [1,2,3]. Additional information reveals that the retention of rigid particles of known size is also not consistent for all 0.2 micrometer filters [4,5]. Thus, it is felt appropriate to review the topic of retention rating and evaluate the applicability of various approaches presented in literature.

VISCOUS-AVERAGE-FLOW PORE

The flow of fluids through a porous filter medium is governed by the same principles involved in the flow of fluid through a pipe. Therefore, the average velocity for laminar tube flow can be adapted to estimate the average pore size of a filter from porosity and permeability data as described in the corresponding ASTM Standard Practices (ASTM Designation: F-902). In essence, the average pore diameter, D, for a sample of flat filter medium can be estimated from

$$(\bar{D})^2 = \frac{Q\ T\ 32\ \eta\ z}{\Delta\ P\ \varepsilon\ A} \qquad (1)$$

Where

Q = volumetric flow rate of the fluid, m^3/S

T = tortuosity factor

η = viscosity of the fluid, Pa s

z = thickness of the medium, m

ΔP = pressure drop across the medium, Pa

ε = porosity of the medium (fractional void volume)

A = filter media surface area, m^2

$\bar{D}$ = Viscous-flow-average pore diameter, m

Various reports in literature indicate that for the common membrane media, the distribution of pores follow a log-normal pattern. The geometric standard deviation depends somewhat on the porosity of the medium, but as reported, a typical value of two can be used to estimate the distribution [6,7].

WET FLOW POROSIMETRY

Microporous membranes may be characterized by wet-flow porosimetry as described in the corresponding ASTM Standard Method of test (ASTM Designation: F-316). This method is based on the fact that porous structures are essentially impermeable to bulk flow of a test gas until a pressure high enough to dislodge the liquid from the pores is reached. This pressure is generally referred to as the bubble point, and is the topic of several articles published in recent literature [8,9,10].

A simple mathematical model to describe the pores of a filter is a manifold of capillaries. Although somewhat unrealistic due to the sponge like reticular structure of most filter membranes, the capillary model lends itself for a theoretical determination of pore sizes. A balance of axial forces involved, i.e., capillary force to retain the water and pressure to displace it from the pore, reveals that d, the diameter of the pore, can be calculated from

$$(P_t) \text{ Pressure} = \frac{\text{Perimeter force}}{\text{Area}} = \frac{\pi \, d \, \gamma \cos\theta}{\frac{1}{4} \pi \, d^2}$$

from which

$$(P_t) = \frac{4 \gamma \cos\theta}{d} \quad (2)$$

Where

P_t = applied test pressure at bubble point, (dynes/cm^2)

γ = the surface tension of the liquid, (dynes/cm)

d = the diameter of the pore, (cm)

θ = the contact wetting angle (the cosine of which is 1.0 if wetting is complete between the liquid and the capillary wall)

Equation 2 holds only for cylindrical capillaries that may not normally be found in common filters. The effect of extreme shape deviation can be illustrated by considering a slit of length L, far larger than its width, W. The balance of forces can be expressed as:

$$2(L + W) \cos\theta = LW \, P_t$$

This reduces to:

$$P_t = \frac{2 \gamma \cos\theta}{W} \quad (3)$$

indicating that at the same pressure, a slit of width only one-half of the diameter of a round pore will be blown open. Most real values will likely fall between these extremes, presumably closer to the capillary than the slit. In addition, it is difficult to measure the contact angle between the liquid and the pore wall directly. A numerical value for the cos θ can be arrived at indirectly by measuring the bubble point of a given membrane in fluids of various surface tensions, and clearing the value from

$$d = \frac{4 \ \gamma 1 \ \cos \theta 1}{P1} = \frac{4 \ \gamma 1 \cos \theta 2}{P2} \qquad (4)$$

where one of the two liquids is known to wet the sample perfectly. By such analysis, it can be shown that relatively hydrophobic materials such as polysulfone have a cos θ value of 0.6 when tested in water. Table 1 lists various test pressures and the approximate corresponding pore diameter calculated from equations 2 and 3.

Table 1 - Pore Diameter (μm) Corresponding to Various Test Pressures

Pressure (psig)	Capillary (cos θ = 1)	Capillary (cos θ = 0.5)	Slit (cos θ = 1)	Slit (cos θ = 0.5)
200	0.2	0.1	0.1	0.05
100	0.4	0.2	0.2	0.1
50	0.8	0.4	0.4	0.2
40	1.0	0.5	0.5	0.25
20	2.0	1.0	1.0	0.5
10	4.0	2.0	2.0	1.0
5	8.0	4.0	4.0	2.0

1 psig = 6.76 kPa

From this Table, it can be surmised that an observed bubble point pressure of 50 psig can indeed correspond to a perfectly wet pore of 0.8 micrometers in diameter, a poorly wet slit of 0.2 micrometers in width or anything in between. It should be obvious that the bubble point alone does not provide sufficient information until a correlation to actual retention data is established. In addition, the bubble point only gives an indication of the size of the largest pore or set of pores, but provides no information about the size distribution. This information has to be obtained by quantitative measurements of the air flow at higher pressures, and relating the air flow through the wet filter to the air flow through the filter at the pressure prior to wetting the filter. The procedure is thoroughly explained in ASTM Designation: F-316.

For the above discussion, it has been assumed that the air flow through the wet filter is zero below the bubble point, but it has been shown that diffusion below as well as above this pressure must be taken into account if highly accurate results are desired [11,12]. Monitoring the transition from diffusive to bulk flow of air through the wet filter will also aid in establishing the intrinsic bubble point of the material being tested. Mathematically, the diffusion through the wet filter as described by Ficks law, mathematical treatment of which is presented elsewhere [12].

PARTICLE RETENTION

It has long been accepted that the best way to determine a filter's retention characteristics is by subjecting it to actual use and monitoring the effluent for the absence of contaminants known to be in the feed stream. While this observation may be fine for selection of filters for a given application, it has also been recognized that model contaminants are a practical approach to describe the filter's retention characteristics in more general terms. Since the retention is, within reason, a reflection of the pore size distribution, retention data of known size contaminants can be used indirectly to estimate the pore size distribution.

Bacteria by far are not the only items critical industries wish to remove from their process solutions. Therefore, other models may well be more suitable to evaluate filter performance. Fine particles of activated charcoal, for instance, are not only often an undesirable "ingredient" present in dechlorinated water or products decolorized by this adsorbent, but also presents a very effective means of comparatively testing membrane filters. If retained on a white membrane, carbon fines will turn it distinguishably black after as little as 0.0005 mg per square centimeter accumulates [5] .

Compared to other contaminants such as AC Fine Test Dust, Latex spheres serve as an excellent model for retention experiments due to their well defined shape. Latex spheres are available in a wide selection of traceable sizes, and generally exhibit a very narrow size distribution. The standard deviation is often less than 2% from the mean, making Latex, for all practical purposes, a mono-sized model. Compared to cumbersome and slow bacteria challenge tests, Latex experiments are easily conducted, as has been shown in literature [13, 14]. By either using a suitable turbidimeter or even visual evaluation against known Latex concentrations, the retention results can be quantitatively evaluated by comparison of the optical density of the particle suspension before and after filtration. A sensitivity range of 5 orders of magnitude, i.e., 0 to 99.999% retention can be covered by this method.

In the case of latex spheres, large amounts of contaminant are necessary in order to detect passage. As many as one million 0.2 micron particles per milliliter may barely be detected by the naked eye or a reasonably sensitive turbidity monitor. Thus, starting with a seemingly unrealistic concentration of 10 billion particles will allow for easy detection of a 10,000 fold reduction in the concentration of the contaminant. If such a reduction is indeed observed, it can be interpreted as the "non-filtration" of one out of every 10,000 units of volume filtered for mechanically retentive filters. This ratio should be independent of concentration. Thus, a filter that reduced the concentration from 10 billion to one million should by rights be able to reach a reduction to 0.01 if the starting concentration was only 100 instead of 10 billion.

By challenging a membrane filter with various sizes of mono sized spheres, information can be gathered and applied in various ways. If the percent retention is plotted as a function of particle size, the graph reflects the filter's pore size distribution. The data can be used if needed to extrapolate into either extreme of retention. Plots of log efficiency versus size or log-log efficiency versus log size are also useful ways of presenting data, particularly when data covering several orders of magnitude needs to be plotted [6].

EXPERIMENTAL PROCEDURES AND RESULTS

Three different membrane types with designated pore ratings of 0.2 micrometers were tested to see how they fit the aforementioned models. Procedures were applied for determining the viscous-flow-average pore diameter in filter media based on ASTM Designation: F316-70. Since there is no standard test methods at the present for evaluating retention efficiencies using latex particles, procedures described were adapted from published literature [5,13,14].

VISCOUS-FLOW-AVERAGE PORE DIAMETER

Air flow rate versus pressure was measured for the membrane samples and the average pore diameter was calculated based on Eq (1). The air pressure was measured with a water filled U-tube manometer for membrane disc samples having a surface area of 47 cm^2 (4.7×10^{-4} m^2) at an air flow rate of 15 SCFH (1.18×10^{-4} m^3/s). The air flow was measured with a Dwyer Rotameter having a range of 0-30 SCFH. The bulk porosity or void volume of the membrane sample was determined by measuring the membrane thickness with a snap gauge micrometer, calculating the total dry volume the membrane occupies and determining the void volume by liquid flooding measurements. The tortuosity factor was taken to be $1/\varepsilon$ or the inverse of the fractional void volume. The results are listed in Table 2.

Table 2 - Terms for Average Pore Diameter Equation

	Membrane Type		
Characteristic	Asymmetric Polysulfone	Nylon 66	Polyvinyldene-difluoride
flow rate, Q (m^3/s)	1.18×10^{-4}	1.18×10^{-4}	1.18×10^{-4}
differential pressure, ΔP (Pa)	1.75×10^{3}	4.125×10^{3}	6.6×10^{3}
thickness, z (m)	1.45×10^{-4}	1.36×10^{-4}	1.09×10^{-4}
fractional void volume, ε	0.85	0.89	0.83
tortuosity factor, $1/\varepsilon$	1.18	1.12	1.20
air viscosity, η (Pa∘s)	1.8×10^{-5}	1.8×10^{-5}	1.8×10^{-5}
membrane area, A(m^2)	4.7×10^{-3}	4.7×10^{-3}	4.7×10^{-3}
V-F-A PD [a] (m)	1.28×10^{-6}	7.7×10^{-7}	5.88×10^{-7}
V-F-A PD (um)	1.28	0.77	0.58

a=viscous-flow-average pore diameter

MAXIMUM AND MEAN FLOW PORE

A 25 mm membrane filter was placed in a stainless steel filter holder and the nitrogen gas flow rate was measured as a function of pressure using a set of rotameters covering the range of 0-30 SCFH (2.36 x 10^{-4} m^3/s). The data was recorded and plotted as depicted in Figure 1. The filter was removed and wet in deionized water replaced in the holder and the gas pressure was increased slowly and incrementally. The pressure was recorded at the first steady stream of bubbles from the outlet of the filter holder fitted with a 1.0 mm I.D. tube. The filter was carefully removed, dried, rewet in mineral oil and installed in the holder and the bubble point was measured as previously described. The 1.0 mm I.D. tube was removed from the holder outlet and the gas pressure was increased while measuring the gas flow rate through the filter using a rotameter downstream of the filter holder. The bubble point pressure values listed in Table 3 are representative of the membrane based on an average of five samples for each type of membrane tested. The mean flow pores pressures listed in Table 3 were determined from the graphs of dry and wet air flow versus pressure for the membrane samples as depicted in figures 1a, 1b and 1c. The corresponding pore diameters also listed in Table 3 were determined from the equations given in ASTM Designation F-316.

Table 3 - Bubble Point and Mean Flow Pore Pressures and Corresponding Pore Diameters (1 psig = 6.76 kPa)

Membrane Type	APS	N66	PVDF
Bubble Point H_2O, psig	53.0	51.0	56.0
Max. Pore Diameter, H_2O, μm	0.6	0.59	0.54
Bubble Point Mineral Oil, psig	23.0	19.0	20.0
Max. Pore Diameter, Mineral Oil, μm	0.63	0.76	0.72
Mean Flow Pore Pressure Mineral Oil, psig	145.0	100.0	80.0
Mean Flow Pore Diameter, μm	0.099	0.144	0.18

LATEX PARTICLE RETENTION

Three (3) different types of membrane discs were challenged with various size latex spheres secured from Duke Scientific using the following procedure: For each of the membrane types tested, several 25 mm discs were cut from a 90 mm and each challenged with a different size latex particle. The discs were cut adjacent to the samples used in the bubble point procedure. The challenge suspension contained 0.1% latex solids concentration and 0.1% Triton X-100 in filtered deionized water.

The filter sample to be tested was installed in a 25 mm filter holder with a volume holding capacity of 15 ml and a vessel outlet fitted with one mm I.D. tube. After wetting the filter samples in deionized water and verifying integrity by a visual bubble point the downstream outlet tubing was drained and the vessel filed with

Figure 1 - Wet-Flow Porosimetry Profiles of Asymmetric Polysulfone, Nylon 66 and Polyvinylidene-Difluoride membranes.

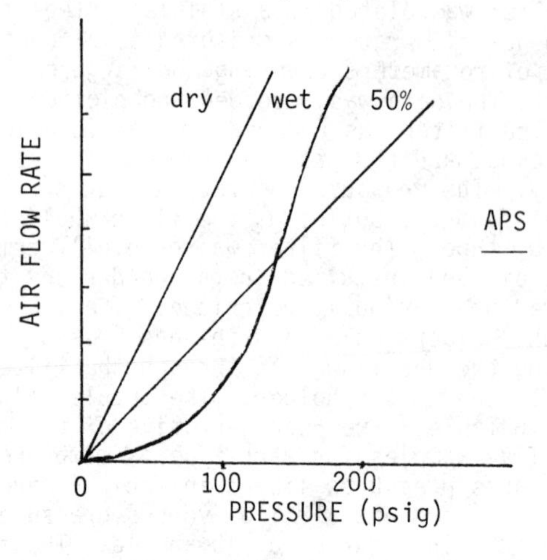

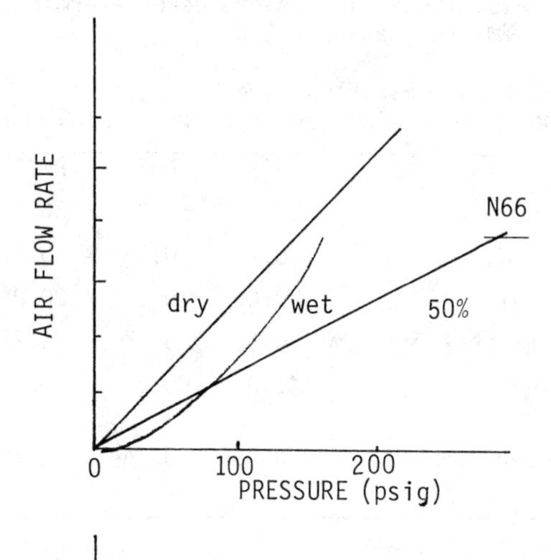

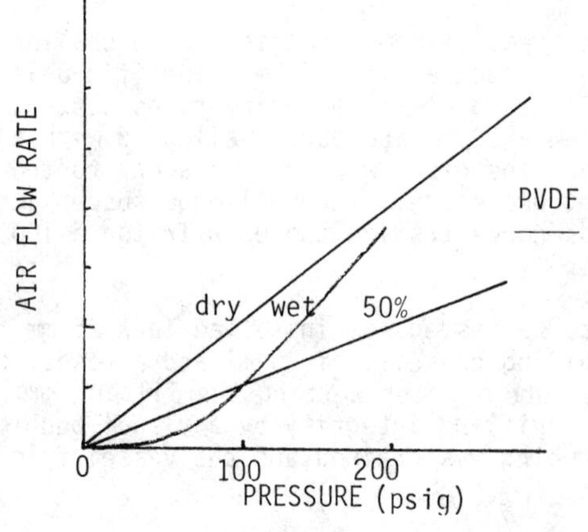

10 ml of the appropriate size latex suspension. The filter holder was then pressurized to 10 psig air pressure and the entire filtrate collected and evaluated for optical density at 420 nm on a Perkin-Elmer Coleman 55 spectrophotometer. This procedure was repeated for all the filter types using various size latex particle. The filtrate and concentrate optical densities were used to calculate the percent retention listed in Table 4.

Table 4 - Latex Particle Retention Results

Particle Diameter, μm	Retention (%) APS	N66	PVDF
0.091	35.0	0	11.0
0.198	98.5	5.1	25.8
0.305	99.7	80.4	75.0
0.460	99.9	99.9	99.9

The previous data was based entirely on experimental procedures using membrane filter discs. To show the comparison between particle retention of 25 mm membrane discs with 3.8 cm^2 filter surface area and cartridge filters having an excess of one thousand times the surface area of a membrane disc (depending on the manufacturer), it was decided to perform retention tests using various size latex particles for 0.2 micrometer rated filter cartridges containing the same types of filter media used in the disc challenge experiments.

The following procedure was used for filter cartridges after rinsing and integrity testing the filter cartridge, four liters of a solution containing 0.091 μm diameter latex particles were passed through the cartridge at a flow rate of two liters per minute into a glass vessel. Samples of the test solution and filtrate solution were evaluated for optical density at 420 nm on a Perken Elmer Coleman 55 spectrophotometer.

The cartridges were then flushed for 5-10 minutes with filtered deionized water until the filtrate was visually clear and then ten liters of filtered deionized water containing 0.1% Triton X-100 was passed through the cartridge into a precleaned glass receiving vessel. This filtrate was analyzed for optical density to serve as a background to demonstrate that the previously run particles were no longer coming through the cartridge. At this point, the system was drained of liquid, and a solution containing the next larger size particle was passed through the cartridge. The solution samples were evaluated for turbidity and this rinse and background determination was repeated until all the sizes of latex spheres mentioned had been filtered through the test cartridge.

The filtrate and concentrate optical densities were used to calculate the percent retention listed in Table 5.

Table 5 - Retention of Various Size Latex Particles By 0.2 μm Rated Filters

Latex Particle Diameter (μm)	Retention (%)		
	Asymmetric Polysulfone	Nylon	Polyvinyldene Difluoride
0.1	6.9	7.9	6.5
0.2	96.0	16.5	44.1
0.3	97.8	49.8	93.6
0.4	99.5	85.4	94.5
0.5	99.9	95.8	97.0
0.6	99.9	97.8	99.1

The latex particle retention results from Table 5 were plotted against the particle diameter and from this graph, the particle diameter at which 50% retention would occur was estimated. The calculated pore diameters for the various pore size determination methods are listed in Table 6.

Table 6 - Summary of Pore Diameters Calculated Based On Several Different Methods

Method	APS	Nylon	PVDF
Viscous-flow-average pore diameter, μm	1.28	0.77	0.58
Mean Flow pore diameter, μm	0.099	0.180	0.144
50% Latex Particle retention pore diameter, μm	0.15	0.030	0.33
Pseudomonas diminuta retention pore diameter, μm	0.2	0.2	0.2

DISCUSSION

As can be seen from the results listed in Table 6, none of the models currently used to describe the pore size distribution of filter media appears to agree with the commonly accepted rating based on organism challenge tests. To a certain extent the discrepancies can be justified. For instance, assumptions have to be made in calculating the viscous-flow-average pore, and the model itself is rather dependant on flow performance characteristics. Yet, it can be shown that flow performance is not drastically affected by the presence of small defects, which obviously could have a detrimental effect on the retention characteristics. Flow rate alone can therefore not be used to predict the outcome of a given filtration.

This can be seen when "mis-applying" the model to highly asymmetric membranes. Although tighter in retentive capability, the viscous flow model predicted a pore size two-fold that of the conventional isotropic membranes included in the study. In asymmetric membranes, the actual distance over which the flow is constricted is far less than the membrane thickness, but even if one tenth of the thickness is assumed and the tortuosity ignored (T = 1.0), the predicted mean pore value would be about 0.3 micrometers, still large compared to the performance observed against known size particle challenges. While this model may be applicable to coarser media such as felts or wound string, it appears the flow performance of thinner media such as the microporous membranes described is better than one would anticipate, thus the resulting rating is somewhat conservative when compared to the retention performance observed.

Assumptions also have to be made in the wet flow porosimetry approach. For example, no attempt was made to correct for the fact that polysulfone wets poorly. The four-fold spread in size between a perfectly wet capillary and a poorly wet slit discussed (see Table 3) may be an invalid extreme, but it would take approximately that factor to make the largest pore (ASTM Designation: F-316) agree with the rating. Surprisingly, the mean flow pore as determined by ASTM Designation: F-316 comes close to the rating, but seldom do users of membranes settle for retaining as little as 50% In addition, it has to be said that the experimentation required for determination of the mean flow pore is somewhat more complicated than what would be anticipated from reading or understanding the theory behind it. It is really not practical at all to try to extend the determination to a pressure at which one-half of the dry air flow is reached, since the flows involved for a 0.5 m^2 cartridge for instance are rather substantial in magnitude ($\sim 10^8$ ml/min. or 1.67 m^3/sec) at the pressures common to such tests. Although the bubble point test has proven to be a practical indicator of filter integrity, it is not really a good indicator for retention of particles smaller than the size which corresponds to the test pressure. This is not only due to the assumptions discussed earlier, but it has been recognized that in an integrity test setting,gas is able to "leisurely" find the largest outlet pore of the individual polyhedral open walled cells that make up the reticulated structures of membrane filters produced by conventional casting techniques [15]. These pores are part of a "maze path" that can turn in any direction, even back toward the pressurized upstream of the membrane. When in use, however, the liquid filtered is physically driven along the shorter, far more direct "tortuous path" across the membrane structure.

Considering that there are inherent weaknesses in the various mathematical approaches one would hope that some agreement can be observed in the rating based on challenge studies. This is indeed the case, since 0.45 micrometer and 0.2 micrometer rated filters repeatedly achieve high levels of retention of *S. marcescens* and *P. diminuta*, the agreed upon organisms. The problem is that *S. marcescens* is not 0.45 nor *P. diminuta* 0.2 micrometers in size. All that these tests measure, in fact, are the degree of perfection to which membrane filter devices can be constructed. Removal efficiencies of 99.999 999 999% are commonly verifiable in short term challenge studies, but as indicated before, performance differences have been reported in long-term studies as well as particle retention experiments. The models currently used to describe the pore size distribution of filter media have certainly been applied in several publications as described earlier in this paper. The data obtained in this experiment is in agreement with data published in literature, and really only indicates once again that microbial 0.2 micrometer rating is not all compatible with the observed particle retention reported in Tables 4 and 5.

It has long been recognized that when a particle confronts a pore of larger size, the particle does not necessarily emerge with the effluent downstream, since adsorptive and depth type of phenomena are invariably involved in the capture. The contrary is true, however: if a particle passes a filter it found a pore large enough to do so. Therefore, an idea of the pore size distribution can be arrived at indirectly by observing the passage of known size particles through filters. While it has to be recognized that various types of media, such as charge membrane for instance, may behave differently and therefore may require special considerations, it is the hope of the authors that the known size particle challenge approach presented can serve as a basis for a review of how filters are rated.

REFERENCES

[1] Simonetti, J.A. and Schroeder, H.G., "Evaluation of Bacterial Grow Through," Journal of Environmental Science, Vol. 27, No. 6, 1984, p. 27.

[2] Howard, G. and Duberstein, R., "A Case of Penetration of 0.2 μm Rated Membrane Filters by Bacteria," Parenteral Drug Association Paper, November, 1979.

[3] Wallhausser, K.H., "Grow Through and Blow Through Effects in Long Term Sterilization Processes," Die Pharmazeutische Industrie, Vol. 45, No. 5, 1983, p. 527.

[4] Wrasidlo, W., Mysels, K.J., "The Structure and Some Properties of Graded Highly Asymmetrical Porous Membranes," Journal of Parenteral Science and Technology, Vol. 38, No. 1, 1984, p. 24

[5] Tolliver, D.L. and Schroeder, H.G., "Particle Control in Semiconductor Process Streams," Microcontamination, Vol. 1, No. 1, 1983, p. 34.

[6] Johnston, P.R., "Fluid Filter Media: Measuring the Average Pore Size and the Pore-Size Distribution, and Correlation with Results of Filtration Tests," Journal of Testing and Evaluation, JTEVA, Vol. 13, No. 4, 1985, p. 308.

[7] Gelman, C., Korin, A., and Meltzer, T., "Development of a New Membrane Prefilter utilizing Pore Size Distribution Analysis," World Filtration Congress III, 1982.

[8] Johnston, P.R., Lukaszewicz, R.C., and Meltzer, T.H., "Certain Imprecisions in the Bubble Point Measurement," Journal of Parenteral Science and Technology, Vol. 35, No. 1, 1981, p. 36.

[9] Reti, A.R., "An Assessment of Test Criteria in Evaluating the Performance of Integrity of Sterilizing Filters," Bulletin of the Parenteral Drug Association, Vol. 34, No. 4, p. 187.

[10] Pall, D.B., and Kirnbauer, E.A., "Bacteria Removal Prediction in Membrane Filters," 52nd Colloid and Surface Symposium, University of Tennessee, Knoxville, Tennessee, June 1978.

[11] Schroeder, H.G. and DeLuca, P.P., "A Rational Approach to Integrity Testing of Filter Cartridges," Interphex 80, Section Research and Development, Sept. 15-17, 1980, p. 62.

[12] Schroeder, H.G. and DeLuca, P.P., "Theoretical Aspects of Sterile Filtration and Integrity Testing," Pharmaceutical Technology, Vol. 4, No. 11, 1981, p. 80.

[13] Pall, D.B., Kirnbauer, E.A., and Barrington, T.A., "Particulate Retention by Bacteria Retentive Membrane Filters," Colloids and Surfaces, Vol. 1, 1980, p. 235.

[14] Simonetti, J.A., and Cooke, M., "State of the Art Filter Pore Size Rating Using Polystyrene Latex Spheres," Presentation to the Society of Manufacturing Engineers, Itasca, Illinois, March 18, 1980.

[15] Williams, R.E., and Meltzer, T.H., "Membrane Structure, the Bubble Point and Particle Retention: A New Theory," Pharmaceutical Technology, Vol. 7, May 1983, p. 36.

Jacqueline P. Bower

CORRELATION OF BIOLOGIC RETENTION, LATEX PARTICLE RETENTION AND PHYSICAL TESTS ON 0.1 MICRON PORE RATED MEMBRANE FILTERS

REFERENCE: Bower, J. P., "Correlation of Biologic Retention, Latex Particle Retention and Physical Tests on 0.1 Micron Pore Rated Membrane Filters," Fluid Filtration: Liquid, Volume II, ASTM STP 975, P. R. Johnston and H.G. Schroeder, Eds., American Society for Testing and Materials, Philadelphia, 1986.

ABSTRACT: A test method is proposed for characterizing 0.1µm-rated membrane using *Acholeplasma laidlawii*, an organism that will pass unimpeded through a 0.2 micron pore size membrane. Membranes were defined physically by bubble point, flow rate, burst strength, weight, air permeability and thickness. In addition, latex particle challenges were performed in an effort to correlate biologic retention and latex particle challenge to the various physical tests performed.

KEYWORDS: biologic retention, latex particle retention, membrane filter, *Acholeplasma laidlawii*

INTRODUCTION

Membrane filters are thin plastic films that are frequently used to stabilize liquids microbiologically. From the user standpoint, the choice of which pore size of membranes to use to achieve microbiological stability requires the end user first define the bioburden or to identify the critical destabilizing organism that must be removed.

When microbiological stability is defined as a sterilizing filtration, as in the pharmaceutical industry, or when *Pseudomonas sp* are known to be present, a 0.2µm-rated filter is selected. Under specific situations, however, such as serum filtration, the smallest destabilizing agent is not a "classic" bacterium per se, rather the smallest organisms present are members of the genera *Mycoplasma* and/or *Acholeplasma*.

Ms. Bower is Director of Technical Services at Micro Filtration Systems, 6800 Sierra Court, Dublin, CA 94568.

From the manufacturer's standpoint, the need to corroborate physical testing of membrane filters with biologic retention was first demonstrated by Bowman et al in 1967[1]. Here, Bowman demonstrated that a small pseudomonad would repeatedly pass through a 0.45μm-rated membrane. This organism was identified as Pseudomonas diminuta and is now the classic challenge organism for testing 0.2μm-rated membrane. Rodgers and Rossmore[2] continued work in this area and proposed methods to qualify membrane pore size by biologic retention in addition to physical methods. The results of their work to characterize membranes by this method for 0.2μm-rated membrane and larger, has generally been adopted by manufacturers as routine quality control protocol.

At Micro Filtration Systems (MFS) we have adopted the following challenge organisms for challenge testing to confirm biologic pore size in the range of 0.2 to 0.8 micron.

0.65 and 0.8 micron	Saccharomyces cereviseae
0.45	Serratia marcescens
0.30	Pseudomonas aeruginosa
0.20	Pseudomonas diminuta

Since this initial work in the late 60's and early 70's, the need to define a biologic pore size for membrane filters smaller than 0.2μm, which are known to pass Mycoplasma and Acholeplasma, has been identified. Methods have been developed at MFS to define a biologic pore-retention test for Acholeplasma sp using 0.1μm-rated pore membrane filters based on ASTM F838-83, Standard Test Method for Determining Bacterial Retention of Membrane Filters Utilized for Liquid Filtrations [3] and HIMA Document No. 3, Vol. 4 "Microbiological Evaluation of Filters for Sterilizing Liquids" [4, 5].

Contaminants that are frequently encountered in serum that has been filtered using a 0.2μm-rated membrane belong to the order MYCOPLASMATALES. This order of microorganisms is comprised of two families, the Mycoplasmataceae which require sterol for growth and the Acholeplasmataceae which do not require sterol for growth. Each family has one genus, the Mycoplasma and the Acholeplasma, respectively. Both Mycoplasma and Acholeplasma are highly pleomorphic, facultatively anaerobic organisms that vary in shape from spherical to slightly ovoid. These organisms lack a true cell wall and are bounded by a single triple layered membrane. Some have postulated that because these organisms lack a rigid cell wall, they do not behave like rigid particles on a membrane surface. Hence, since they may be considered deformable particles, this accounts for their ability to pass through 0.2 micron pore size membranes. Cells stain Gram negative and range in size from 0.1 to 0.25 micron in size [6].

In our studies we chose an organism that was isolated from nonsterile serum that had been filtered using a 0.2μm-rated cellulose nitrate membrane. This contaminant has been identified as Acholeplasma laidlawii.

EXPERIMENTAL METHOD: BIOLOGIC RETENTION TEST

Test Procedure

Medium: Mycoplasma (PPLO) broth was prepared as directed by the manufacturer.

90ml of Mycoplasma Broth (BBL), 0.5g yeast extract and 0.2ml of 1% phenol red solution were mixed and sterilized by ethylene oxide gas. Horse serum (10ml), 100,000 IU Penicillin G and 1.5ml 50% glucose were each filter sterilized and introduced to the solution as sterile additives.

Innoculum: Freshly prepared medium was innoculated with *Acholeplasma laidlawii* and cultured anaerobically at 37°C for 96-120 hours. Cell concentration was determined by preparing serial dilutions and culturing anaerobically on Mycoplasma agar at 37°C for 96-120 hours.

Positive Controls: Serial dilutions of the innoculum were prepared to yield a final concentration of approximately° 10 organisms per milliliter of medium. Diluted innoculum is added to the fresh medium and cultured at 37°C for 21 days. A low level positive control was adopted because we felt it to be a more severe, representative test of actual passage of *Mycoplasma sp* through a membrane filter. In addition, we felt it necessary to validate the ability of the medium to detect minimal passage of the challenge organism through the membrane.

Negative Control: Prepared flasks of fresh uninnoculated medium were cultured anaerobically at 37°C for 21 days.

Test Equipment

Five 0.1μm-rated pure nitro cellulose, 47mm diameter membrane filters or 62.5cm^2 of effective filter area are tested as replicates at each sampling point on the roll of membrane filter. These discs were loaded into each of 5 MFS Model KST47 (Catalog Number 301500) stainless steel filter holders. Outlet connectors are wrapped with Kraft paper and the entire assembly is autoclaved at 121°C for 20 minutes, slow exhaust. Holders are allowed to cool then they are mounted vertically.

Innoculum is measured from a large Erlenmeyer culture flask into a sterile graduated cylinder and added to the filter holder reservoir. The culture flask is kept on a magnetic stir plate with constant stirring between tests to assure that the organisms are monodisperse. Innoculum volumes are adjusted to yield approximately a monolayer challenge of 10^7 organisms per square centimeter of filtration area.

Units are sealed and pressurized using nitrogen gas to 0.5kg/cm^2 and held for 5 minutes. Serum, being a complex biological fluid is subject to protein denaturation and formation of a gel layer on the membrane filter surface when differential pressures exceed 0.7kg/cm^2 Typically, serum filtrations are performed at 0.1-0.4 kg/cm^2. Therefore, selecting a test differential pressure of 0.5kg/cm^2 represents an extreme case and more severe pressure than would normally be encountered. Filtrate from each holder is collected into individual flasks containing 50 ml of fresh sterile medium.

After the 5 minute time period has elapsed, the units are rapidly depressurized to atmosphere. After 60 seconds, the units are again repressurized to 0.5kg/cm^2 and held for 5 minutes. This pulsed cyclic operation which simulates a pinch-fill operation is repeated 10 times.

After the filtration cycle is completed, flasks are sealed and cultured anaerobically for 21 days at 37°C. Passage of the challenge organism is noted when the medium changes from the characteristic red color to yellow.

Cell size of the Acholeplasma laidlawii challenge was confirmed using scanning electron microscopy at 20,000X magnification. Cell dimensions were 0.15X0.4 microns.

Air Permeability

Air permeability (AP) measurements were made on all 0.1 micron pore size pure nitrocellulose 47mm diameter membrane filters using a Gurley densometer. Here, air permeability is defined as the time in seconds required to displace 300cc of air through 645.16mm^2 area filter under a cylinder weight of 567 $\pm$1g.

Latex Particle Challenge

Membrane filters which had air permeability values that correlated with minimum Biologic Log Reduction of 7 were selected for further retention tests using latex particle challenge. Individual filters were floated on water to prewet them and promptly transferred to each of 5 MFS Model KST47 (Catalog Number 301500) stainless steel pressure holders. Holders were mounted vertically.

Polystyrene latex particles (Dow Chemical) having a mean diameter of 0.109 microns and having been previously demonstrated to pass quantitatively (LRV=0) through a 0.2µm-rated membrane were used to challenge the filter. Particles (0.1wt%, $1x10^{12}$/ml) were suspended in a 0.3% solution of sodium docecyl sulfate. To minimize aggregates, the solutions were ultrasonically dispersed immediately prior to the test. A challenge volume of 100ml or 10^{11}particles/cm^2 was filtered through the test membrane at 1kg/cm^2 pressure. Filtrate was collected in a clean Erlenmeyer flask. To enumerate the passage of the particles, the filtrate was then filtered under vacuum through a 0.05 micron pore size polycarbonate filter using a MFS Model KG13A filter holder. Captured particles were counted using an electron microscope at 20,000X magnification. Ten to 40 microscopic fields per filter were counted and averaged.

Results

Pure 0.1µm-rated nitrocellulose membrane filters from 3 separate lots of media were selected for these studies.

TABLE 1--
Typical Specifications for 0.1μm-rated
Cellulose Nitrate Membrane

Isopropanol Bubble point (kg/cm^2)[a]	2.9
Water flow rate $(ml/min/cm^2)$[b]	2.7
Burst strength (kg/cm^2)[c]	>1.45
Weight (mg/cm^2)	6.1
Nitrogen flow rate $(l/min/cm^2)$[d]	0.67
Thickness (μm)	110
Porosity (%)	67

[a] The minimum pressure in kg/cm^2 required to force air through a membrane prewet with isopropanol.
[b] Initial flow rate in $ml/min/cm^2$ using D.I. water prefiltered to 0.1 micron at $0.7kg/cm^2$ differential.
[c] Burst strength is the pressure in kg/cm^2 required to rupture a dry unsupported membrane.
[d] Nitrogen flow rate reflects initial flow rates of prefiltered (0.2μm) nitrogen in $l/min/cm^2$ at $0.7kg/cm^2$ differential pressure.

A total of 121 pure nitrocellulose 47mm diameter filters were tested individually, first for air permeability, then subjected to biologic retention tests using *Acholeplasma laidlawii*. Based on these paired tests using these 3 lots, it was determined a minimum air permeability of 210 seconds would give an *Acholeplasma* LRV of at least 7.7. (Table 2)

TABLE 2--
Minimum, Maximum and Mean Air Permeability and
Acholeplasma LRV Reduction Values for
Sterilizing Grade 0.1μm-rated Membrane Filters

	Min	Mean	Maximum
Air Permeability (sec)	210	256	310
Acholeplasma LRV	7.7	8.25	8.8

Only membrane filters which met the minimum air permeability criteria for *Acholeplasma* retention, which had been established in earlier studies [7], were then subjected to the latex particle challenge. Retention of the latex particles is expressed as the Log Reduction Value (LRV). LRV is defined as the logarithm to the base 10 of the ratio of the number of organisms, or particles in this case, in the challenge to the number of organisms in the filtrate. Expressed mathematically,

$$LRV = \frac{\text{Log}_{10}(\text{Total is Number of Latex Particles in the Challenge Filtrate})}{\text{Log}_{10}(\text{Total Number of Latex Particles in the Filtrate}}$$

Paired Air Permeability and Latex Particle Challenge data are summarized in Table 3.

TABLE 3--
Latex Particle Log Reduction Values versus
Air Permeability for 0.1μm Rated Membrane Filters

Trial	Air Permeability (seconds)	LRV
1	226	0.37
2	245	0.41
3	282	0.64
4	288	0.77
5	301	1.1
6	302	1.1

Nitrocellulose membrane filters with similar air permeability values exhibited different LRV's depending on the challenge. In this study, membrane tested with Latex Particle had an LRV of 0.37-1.1. When the challenge was *Acholeplasma laidlawii* the LRV was 7.7-8.8.

Such differences imply that additional forces account for the enhanced removal of *Acholeplasma* compared to latex spheres. Namely, that *Acholeplasma* removal occurs due to adsorption or hydrophobic interaction, whereas latex particle removal is due solely to mechanical separation [8, 9]. It is clear, however, that as the air permeability increases, the retention of latex spheres is similarly increased. Correlation coefficient (r) was calculated at 0.9285 with a y intercept of -1.87 and slope of 0.0095.

Discussion

The definition of a sterilizing grade filter set forth by the Health Industry Manufacturers Association (HIMA) requires that a minimum $1x10^7$ challenge organism per square centimeter of effective filtration area yield a sterile filtrate. Using *Acholeplasma laidlawii*, all membrane tested met this requirement. By changing the challenge however to latex particles, membrane with the same air permeability no longer can be defined as "sterilizing" because the highest LRV observed was 1.1. Differences between the two LRV values can be attributed to either differences in the sensitivities of the two methods, or to differences between two similarly sized but chemically and physically different carriers.

It has previously been reported in the literature [8] that the active retention mechanism for cellulosic membrane filters is due to adsorption. This was also observed in our studies where we see that much higher retentions are observed with *Acholeplasma* as opposed to the latex spheres. Latex spheres are typically covered with surfactant to render them hydrophilic and monodisperse. The surfactant, SDS in this case, also serves to override any irreversible adsorption of the latex particles to the cellulosic matrix. Therefore, any removal of latex particles must be attributed to mechanical separation and not due to adsorption. One additional effect of the surfactant on the bead may be to lower the overall surface tension of the bead thereby rendering it more slippery than a similarly sized microorganism. This slippery effect would not only minimize contact time for adsorption to occur but might also serve to lubricate the particle and cause more passage than would be observed with even a third similarly sized carrier particle.

Previously it was mentioned that it has been postulated that *Acholeplasma*-like organisms pass through filter media because they behave like deformable particles. Our data does not support this. On the contrary, we observed significantly higher retention of *Acholeplasma* compared to rigid latex particles of comparable size. Further, it appears that the more rigid particle (latex) of similar size was able to pass through the media freely, both due to lack of adsorption and lack of mechanical impedance. The mechanical aspects of removal are a focal point of continuing studies where we are coating latex spheres with *Acholeplasma* cell membranes and repeating the latex particle challenge studies.

Conclusions

Two methods for assessing retentive capabilities of 0.1μm rated membranes have been discussed.

A method for assessing biologic retention using *Acholeplasma laidlawii* has been proposed. This method, based on both ASTM and HIMA guidelines can be used to determine the ability of a membrane filter to remove *Acholeplasma* from a liquid sample. Such information or validation of a membrane using this method would be of particular interest to persons routinely filtering sera or persons who are plagued with contamination due to *Mycoplasma*-like organisms. This procedure should only be performed by qualified microbiologists under proper laboratory conditions. Special culturing conditions, namely anaerobiosis, are required. In addition, filtrate must be cultured for a full 21 days before non-passage of organisms can be confirmed.

A second method for assessing the retention capability of 0.1 micron pore size membrane was discussed. This method utilized uniform dispersions of latex spheres. Removal of latex spheres correlated well to air permeability measurements. Log reduction values varied depending upon that challenge organism or particle. Significantly larger LRV was observed using a biologic challenge compared to latex spheres of similar size. The data presented confirms the generally accepted concept that removal of microorganisms using cellulosic membranes is due to adsorption and hydrophobic interactions. Removal of latex spheres however was not due to either of these forces and therefore due only to mechanical separation. Because two entirely different mechanisms are in operation care must be taken when analyzing data from either method and trying to extrapolate to the other method. Specifically, based on our studies, it is premature to assume that an LRV of 1.1 using latex beads, for example, will give an LRV of >8.8 using *Acholeplasma laidlawii*.

ACKNOWLEDGMENTS

The author acknowledges with appreciation the help of Mr. O. Kurisaka and Mr. T. Sakuma in the preparation of this paper.

REFERENCES

[1] Bowman, F.W., Calhoun, M.P., and White, M. "Microbiological Methods for Quality Control of Membrane Filters," J. Pharm. Sci., Vol 56, No.2, 1967 pp. 222-225.

[2] Rogers, B.G. and Rossmore, H.W., "Determination of Membrane Filter Porosity by Microbiological Methods," in Developments in Industrial Microbiology, ed. Corum, C.J., et al Washington, D.C., American Institute of Biological Science, Vol 11, 1970, pp. 453-459.

[3] ASTM F838-83 "Standard Test Method for Determining Bacterial Retention of Membrane Filters Utilized for Liquid Filtration".

[4] "Microbiological Evaluation of Filters for Sterilizing Liquids," HIMA Document No.3 Vol.4 Health Industry Manufacturers Association, Washington, D.C. April 1982.

[5] Leahy, Timothy J., and Sullivan, Mary J., "Validation of Bacterial Retention Capabilities of Membrane Filters", Pharmaceutical Technology, Vol. 2, Nov 1978 pp. 65-75.

[6] Freundt, E. A., "The Mycoplasmas," in Bergey's Manual of Determinative Bacteriology, The Williams and Wilkins Co., Baltimore 1974 pp. 929-955.

[7] Bower, Jacqueline P., "Definition and Testing of a Biologically Retentive 0.1 Micron Pore Size Membrane Filter," presented at Society of Manufacturing Engineers annual meeting March 26-28, 1985.

[8] Tanny, G. B., Strong, D. K., Presswood, W. G., and Meltzer, T. H., "The Adsorptive Retention of Pseudomonas diminuta by Membrane Filters," J. Parent Drug Assoc. Vol. 33, No. 40 1979 pp. 40-51.

[9] Wrasidlo, W. and Mysel, K. J., "The Structure and Some Properties of Graded Highly Asymmetrical Porous Membranes," Parenteral Science and Technology, Vol. 38, No. 1, Jan./Feb. 1984, pp. 24-31.

Peter R. Johnston

ANOTHER VIEW OF THE INTEGRITY TEST, OR BUBBLE-POINT TEST, APPLIED TO FLUID FILTER MEDIA

REFERENCE: Johnston, P.R.,"Another View of the Integrity Test, or Bubble-Point Test, Applied to Fluid Filter Media," Fluid Filtration: Liquid, Volume II, ASTM STP 975, P. R. Johnston, and H. G. Schroeder, Eds., American Society for Testing and Materials, Philadelphia, 1986

ABSTRACT: With the recent realization that the random pore-size distribution in membranes and papers can be described by way of rather simple probability calculations, the principle of the bubble-point test is reviewed to explain previous experimental findings. Discussed here are the meaning of the largest pore, the limit of the bubble-point test to determine its size, and how the perceived bubble point increases with the thickness of the medium.

KEYWORDS: most probable pore-size distribution, average pore size, permeability, bubble point

INTRODUCTION

By integrity is meant that the filter medium (usually a thin material) has no tears or holes and is well sealed into its housing. An integrity test is based on that law of capillary science saying that where a porous material is saturated with a liquid, and a gas is used to blow the liquid from the pores, the pressure required is inversely proportional to the diameter of the pores. Which is to say that a filter assembly will pass an integrity test if the bubble-point pressure reaches a required level.

But, while the principle of the bubble-point test is simple on its face there are complications which have bred confusion among those performing the test, reporting the results, and understanding the results. What follows is a re-examination of the principles offered in a new light.

This new light is the realization that in those filter media with a random distribution of pore sizes we know the shape of that distribution and, as a result, can address the meaning of the "largest" pores.

Johnston is Senior Project Engineer, Ametek, Inc., 502 Indiana Ave., Sheboygan, Wisconsin 53081

Further, we can understand the flow of gas as it displaces liquid from the pores. Part of this understanding enables us to see how the bubble point seems to increase with the thickness of the medium--an observation reported by Pall & Kirnbauer (1) but which has not apparantly been discussed in the literature. Another part enables us to understand the elusive nature of what constitutes the bubble point. To present this discussion it is necessary to devote some page space to a review of the subjects of pore-size distribution and of fluid flow through those pores.

PORE-SIZE DISTRIBUTION

The present discussion is directed to either of three types of filter media: non-woven fibrous materials, membranes manufactured by the solvent-cast, phase-inversion process (the sponge-like membrane), and that singular product, the Nuclepore™ membrane, an otherwise solid sheet of plastic with holes drilled straight through. In the Nuclepore membrane some of these equal-diameter holes touch to yield a larger hole; hence, there exist a random distribution of pore sizes.

Filter media composed of sintered particles or spheres are excluded from this discussion because in those media the pore-size distribution is appartently different (2, 3).

The concept of pore size in the Nuclepore membrane is easily understood. Under the electron microscope one can see neat, circular holes going straight through an otherwise solid sheet. In the case of the sponge-like membrane or a non-woven fibrous sheet the diameters of the pores and the depths of the pores can only be understood from a statistical view. That view addresses the unit depth of a single pore as the thickness of a theoretical plane in the medium, perpendicular to the flow of fluid. This plane, with its "unit" thickness, is composed of a random distrubution of openings of various shapes and cross-sectional areas; and, the medium as a whole consist of a stack of such planes, with some finite distance between each plane. But, for example, a large hole in one plane does not necessarly occur opposite an equally large hole in the adjacent plane.

As a fluid, in viscous flow, passes through a sponge-like membrane or a fibrous sheet it defines an average-size pore. That is, the fluid in passing through one of the theoretical planes defines an average-diameter pore. And, the fluid, in traveling from one face of the medium to the next face, travels an average distance longer than the thickness of the medium by a factor called the tortuosity factor, T, to define the average length of a pore.

The average pore diameter, $\bar{D}$, defined by viscous flow of a particle-free fluid, is calculated (ASTM F902, ref. 4) from measurements in the expression

$$(\bar{D})^2 = \frac{Q\ 32\ n\ z\ T}{A \epsilon\ \Delta P} \qquad (1)$$

where: Q = volumetric flow rate of fluid, m^3/s.
n = absolute viscosity of the fluid, Pa s.
z = thickness of the medium, m
T = tortuosity factor. In the Nuclepore membrane T = 1.0; in the other media addressed here $T = 1/\epsilon$.

ϵ = porosity, ratio: void volume/bulk volume in the medium.
A = area of the face of the medium, m^2
ΔP = pressure difference across the two faces of the medium, Pa

It is fairly well established (4-7) that in the present filter media of interest the random, pore-size distribution closely follows a log-normal distribution; and, the geometric standard deviation (explained below) lies in the range of 1.5 to 2.5--depending on the porosity. That is, the pore-size distribution in a Nuclepore membrane, where porosity is near 0.1, is narrower than in the other media where porosity is near 0.8.

The present discussion--in explaning by example--addresses a medium where $\epsilon = 0.8$, and the geometric standard deviation, explained in Fig.1, is 2.0.

As an aside: It must be understood that while Fig. 1 shows how a stream in viscous flow is distributed among various-size pores, there are two other pore-size distributions of interest. One involves pore volume, reached via mercury-intrusion measurements (that is, pore area with unit depth). That distribution would be shown by a line parallel to the line in Fig. 1 but to the left. The other distribution, also parallel, but even more to the left, would show the number of pores of increasing diameter--such as seen in a microscopic analysis.

Having now addressed how to deduce the flow-average pore diameter via Eq 1 it is of interest to explain where that average value lies in the distribution. To answer that question requires a decision on how much of the distruoution one wants to embrace.That is, in embracing the distribution to $\Sigma Q = 0.99$ the average value of D corresponds to the 0.875 mark on the ΣQ scale. On embracing more of the distribution, the average value of D, as indicated on the ΣQ scale, increases as follows (example in ref. 4):

Fullness of the distribution	Position on ΣQ scale relating to avg. D
0.99	0.875
.999	.90
.999 9	.92
.999 99	.935
.999 999	.948
.999 999 9	.958
.999 999 99	.966

The present discussion assumes $\bar{D}$ corresponds to the 0.95 mark on the ΣQ scale.

AIR PERMEABILITY OF A FILTER MEDIUM

In the present example of a filter, where $\epsilon = 0.8$ and thickness, z, is 130 μm, a measure of air permeability is provided by the data in Fig. 2. The plot in Fig. 2 is the usual way of beginning a study of the extended bubble point or flow-pore distribution (7). And, to reach the size of the average pore, defined by viscous flow of air, one would substitute in Eq 1 values of Q/A and P (that is, ΔP) at the low end of

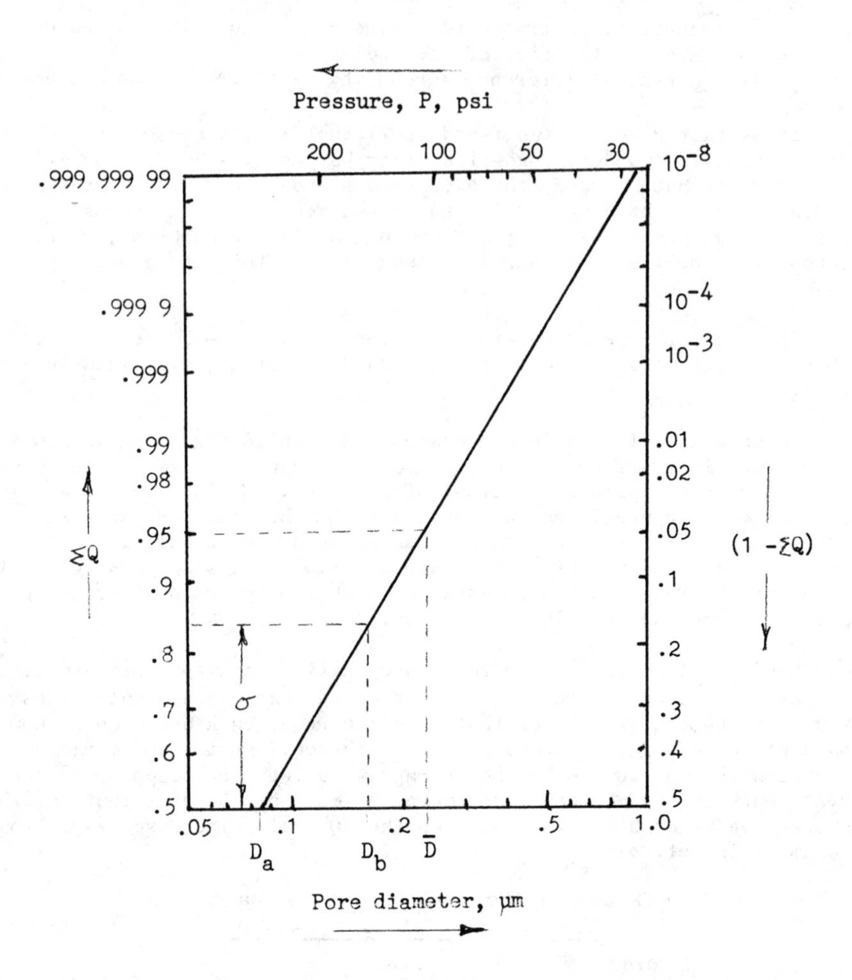

FIG. 1--Top half of the most probable pore-size distribution (Ref. 4). ΣQ = cumulative volumetric flow rate of fluid through pores of diameter D. σ = standard deviation in the normal distribution. P, viewed in relation to $1 - \Sigma Q$, equals air pressure required to blow open the largest, then the next largest pores. See text for method of relating P to D.

In this figure the horizontal axes are laid out on ordinary logarithm scales. The verticle axes are laid out on a scale corresponding to log log $1/(1 - \Sigma Q)$. These verticle axes, when compared to the normal probability scale, are more compressed at the extreme values. A true log-normal shown in this plot would not be straight to the extremes but would curve toward larger values of D. Here the geometric standard deviation is

$$D_b/D_a = 0.165/0.081 = 2.06$$

and $\bar{D}$ is the arithmetic-average pore diameter. 1 psi = 6.895 kPa.

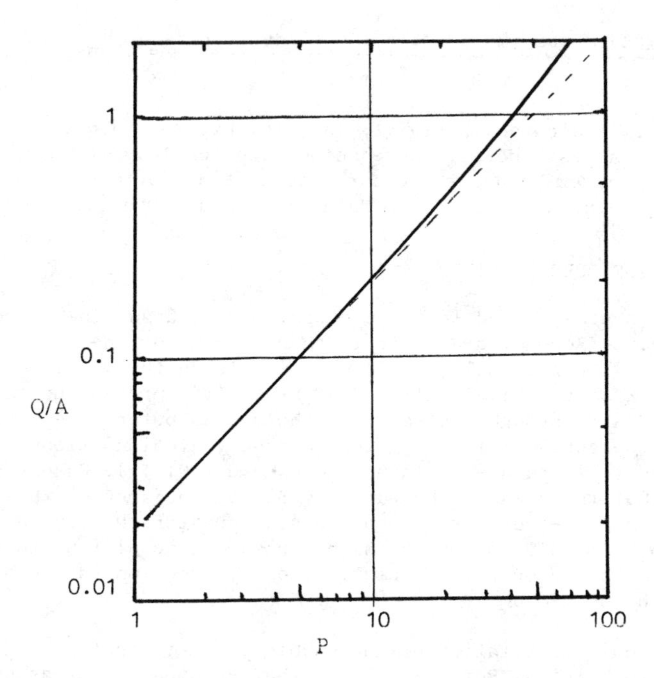

FIG. 2--Air flow rate through the medium of the present example as a function of driving pressure. Q/A = L/min cm^2 emerging from the downstream side at atmospheric pressure (zero gauge pressure). 1 L/min cm^2 = 0.167 m/s. P = gauge pressure on the upstream side, psi. 1 psi = 6.895 kPa.

the scale where it may be assumed that viscous flow does occur. In this example $\bar{D}$ in Eq 1 becomes 0.26 µm, as shown in Fig. 1.

RETURNING TO FIG. 1

The P scale in Fig. 1 is constructed as follows. Assume the medium is saturated with water and then calculate the pressure required to blow open the different diameter pores (assuming the pores are circles) from the expression (8)

$$\text{Pressure} = \frac{\text{Perimeter force}}{\text{Area}} = \frac{2\pi r\ S \cos\theta}{\pi r^2} = \frac{2\ S \cos\theta}{r} = \frac{4\ S \cos\theta}{d} \qquad (2)$$

where: S = surface tension of water in contact with air
θ = angle with which water wets pore walls
r = radius of the pore
d = diameter of the pore

Assuming, in the present example, the wetting angle is zero, so that cos θ = 1.0, and assumimg the surface tension of the water (with some surfactant present) is 46 dynes/cm (9), the pressure required to blow open a 0.9-µm-diameter pore is 28 psi:

$$P = \frac{4 \times 43 \text{ dynes/cm} \times 1.45 \times 10^{-5} \text{psi/dynes/cm}^2}{0.9 \times 10^{-4} \text{cm}} = 28 \text{ psi}$$

which fact is indicated in Fig. 1. That is, Fig. 1 indicates that at 28 psi the "largest" pores, those accounting for 10^{-8} of the fluid flow, are blown open. At 50 psi those pores accounting for 0.5×10^{-3} of the total flow are blown open. These data are now used to construct Fig. 3.

FIG. 3 AND THE BUBBLE POINT

Fig. 3 is constructed in three steps. First, draw Line I from the data in Fig. 2. Second, draw Line II in relation to Line I from the data in Fig. 1. For example, at 50 psi Line II on the verticle scale is 0.5×10^{-3} of Line I. Finally, draw Line III representing the flow of diffussed air through the water in the medium before the water is blown from the pores. This last step is done via calculations that are somewhat lengthly and are explained elsewhere (10, 11). Essentially, one considers that the area of the water is 80% of the area of the medium and that with a pressure of air on one side greater than on the other air will go into solution on the high-pressure side, diffuse to the low-pressure side and come out of solution to give the impression of gaseous flow through the medium.

The logical definitation of the bubble point is that pressure corresponding to Point a. Reti (10) shows that Point a is not as sharply defined as shown here but instead there is a rounded corner. Reti does not define the bubble point but mentions that its pressure corresponds to a flow of air at "several" mL/min since at that flow rate the flow is first perceptable. As Johnston & Meltzer discuss (8) the first perceptable flow depends on the area of medium under study. Consider that one measures the bubble point of a 47-mm-diameter disc on the one hand and a 293-mm disc on the other. Consider further that air emerging from the downstream face of either disc passes through an eye dropper held under water, and that the first perceptable flow of air occurs at a rate of 5 mL/min. Under these conditions, in Fig. 3, the bubble point of the 47-mm disc would be 52 psi, Point c; and, the bubble point of the 293-mm disc would be 38 psi, Point b. Further, if the area of the medium were that in a pleated cartridge--4500 sq cm--the perceived bubble point would be Point d, which point is below the bubble point and in the air-duffussion region.

Pall & Kirnbauer (1) reach a meaning of bubble point via plotting these kind of data in the manner shown in Fig. 4. That is, they plot Q/AP verses P (instead of Q/A verses P) making the plot on linear coordinates, and from the extension of the two essentially straight lines deduce an intersection point corresponding to a pressure they label K_L, which in the present example equals 43 psi.

BUBBLE POINT AND THICKNESS OF THE MEDIUM

Pall & Kirnbauer (1) present data from which one may conclude that the pressure corresponding to K_L increases by a factor of 1.05 every time the thickness of the medium increases by a factor of 2.0. This relationship is shown in the present model by the plots in Fig. 5. Here points r, s & t correspond to the K_L values which do increase with

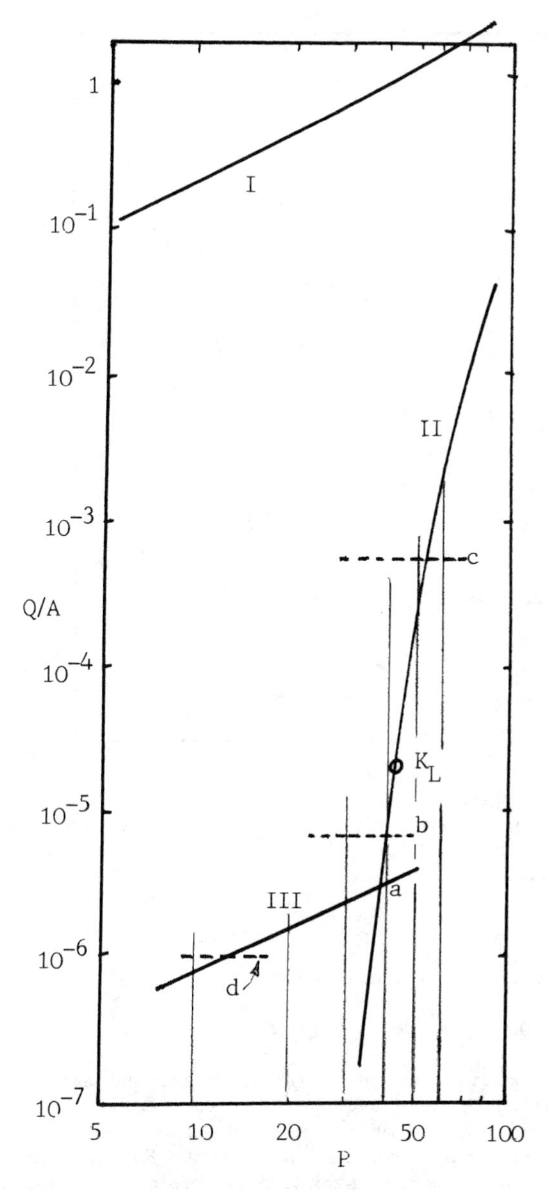

FIG. 3--Definitions of the bubble point (b.p.). Curve I is that curve in Fig. 2. Curve II is derived from Fig. 1, see text. Curve III shows flow of diffused air through the water-soaked medium before pores are blown open. a = a logical definition of the b.p. c = perceived b.p. of a 47-mm-diameter disc, air flow is 5 mL/min. b = perceived b.p. of a 293-mm disc. d = perceived b.p. of a cartridge containing 4500 sq cm of this medium. K_L = Pall-Kirnbauer b.p., explained in Fig. 4.

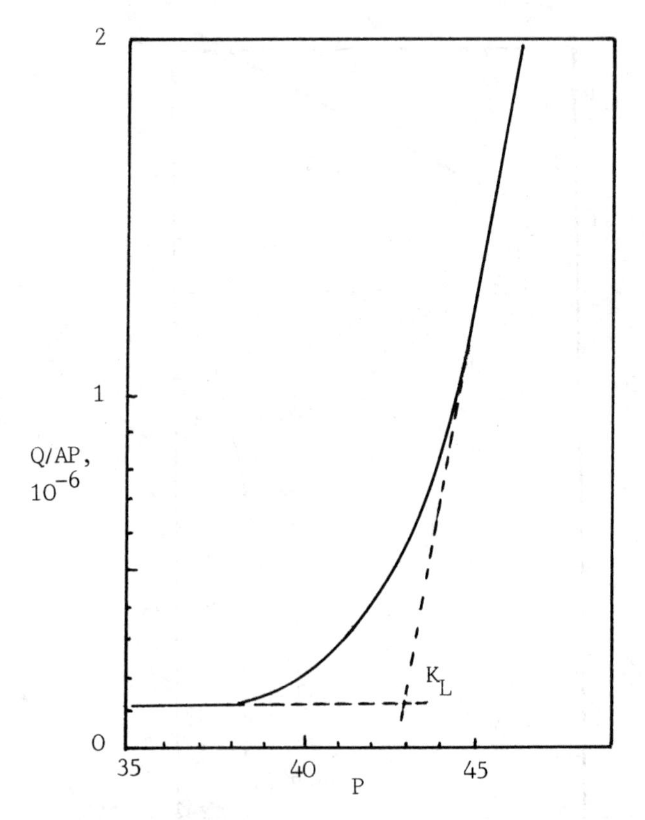

FIG. 4--Pall-Kirnbauer method of reaching the bubble point they label K_L (1). A replot of data in Fig. 3 but now on linear coordinates and with showing Q/AP versus P rather than Q/A versus P.

the thickness of the medium by this factor.

BUBBLE POINT AND POROSITY

Given two filter media with the same thickness and material of construction, and with the same average pore size defined by Eq 1, but with different porosities, the medium with the lower porosity will have the higher value of K_L or perceived bubble point--for two reasons: (a) With lower porosity the geometric standard deviation is less, meaning the pore-size distribution is narrower, hence, there are fewer large pores; (b) Lower porosity means a lower value of Q/A in Figs. 3 and 5 for a given P; hence, a greater value of P is required to see a perceptable Q/A.

CAVEAT

The above directions for constructing Fig. 3 have been simplified so as to show the shape of Curve II in relation to Curve III. More detailed directions would include two additional considerations (7):

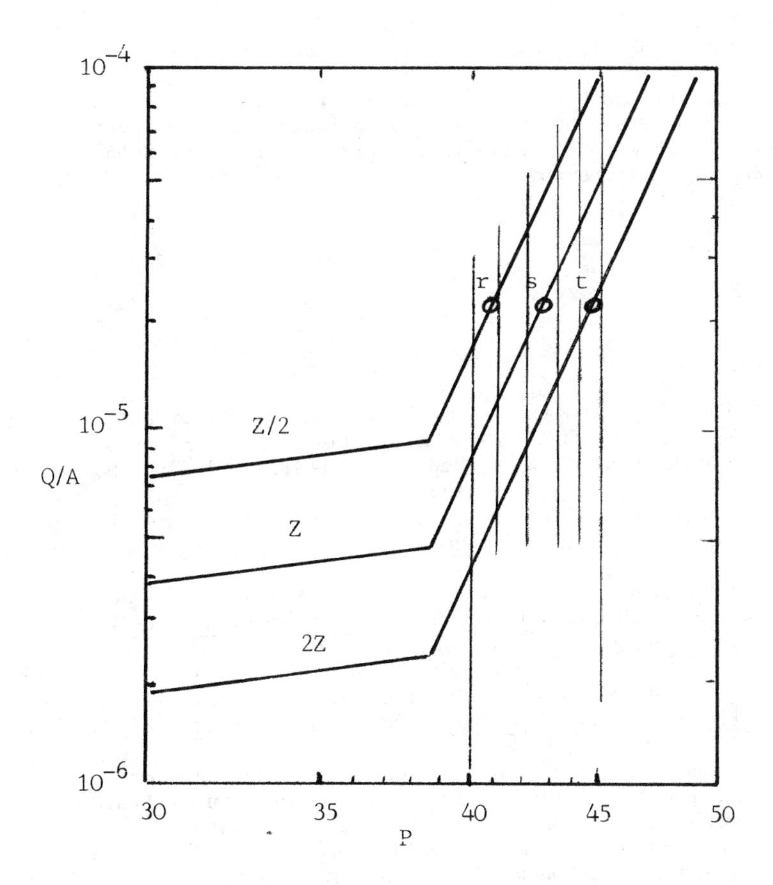

FIG. 5--Illustration of how, in the present model, the Pall-Kirnbauer bubble point, r,s, and t, increases with the thickness of the filter medium. Z = Curves II and III in Fig. 3. Z/2 = results seen where medium is half as thick. 2Z = results seen where medium is twice as thick.

(a) In drawing Curve II in relation to Curve I by the relationship shown in Fig. 1, a better procedure would be to first establish a Curve Ia as frame of reference, rather than Curve I. A Curve Ia is parallel to Curve I but with some lower Q/A values. Curve Ia would represent the flow of air through the medium where even though all the pores have been blown open, the medium is not dry and is therefore not as permeable. In a proper study of the extended bubble point the investigator would wet the medium with a liquid that does not evaporate; or, if an evaporating liquid is used, the air is presaturated with the vapors of that liquid. (b) Where high air pressures are required to achieve flow through the medium the flow of air may not be purely viscous; hence, Curve II would represent a mixture of the viscous-flow-distribution as well as the volume-distribution of pore sizes. Which is to say that Curve II will lean more to the right and (or) be displaced more to the right.

CONCLUSIONS-SUMMARY

An attempt has been made to explain the meaning of--and the different meanings of--bubble point, and to imply that the investigator who reports a bubble-point value is obliged to relate the details of how he made that measurement.

REFERENCES

(1) Pall, D.B. and Kirnbauer, E.A.,"Bacteria Removal Prediction in Membrane Filters," 52nd Colloid Surface Science Symposium, Knoxville, Tenn, 1978. Copy available from Pall Corp., 30 Sea Cliff Ave., Glen Cove, New York 11542.

(2) Grace, H.P.,"Structure and Performance of Filter Media, Part I," American Institute of Chemical Engineers Journal, Vol. 2, No. 3, Sept 1956, pp. 307-315.

(3) Haring, R.E. and Greenkorn, R.A.,"A Statistical Model of a Porous Medium with Nonuniform Pores," American Institute of Chemical Engineers Journal, Vol. 16, No. 3, 1970, pp. 477-483.

(4) Johnston, P.R.,"Fluid Filter Media: Measuring the Average Pore Size and the Pore-Size Distribution, and Correlation with Results of Filtration Tests," Journal of Testing and Evaluation, Vol. 13, No. 4, July 1985, pp. 308-315

(5) Gelman, C., Korin, A. and Meltzer, T,"The Development of a New Membrane Prefilter Utilizing Pore Size Distribution Analysis," World Filtration Congress III, 1982. Copy available from Gelman Sciences, 600 South Wagner Rd., Ann Arbor, Mich. 48106

(6) Kanide, K. and Manabe, S.,"Characterization of Straight-Through Porous Membranes," in Cooper, A.R., Ed., Ultrafiltration Membranes and Applications, Plenum Press, New York, 1980.

(7) Johnston, P.R.,"The Most Probable Pore-Size Distribution in Fluid Filter Media," Journal of Testing and Evaluation, Vol. 11, No. 2, 1983, pp. 117-125.

(8) Johnston, P.R. and Meltzer, T.H.,"Suggested Integrity Testing of Membranes Filters at a Robust Flow of Air," Pharmaceutical Technology, Vol. 4, 1980, pp. 49, 50, 52, 54, 57, 59.

(9) Mehta, D., Hauk, D. and Meltzer, T.,"The Influence of Wetting Agents on the Bubble Point of Membrane Filters," 2nd World Filtration Congress, 1979. Copy availabe from Gelman Sciences, 600 South Wagner Rd., Ann Arbor, Mich. 48106.

(10) Reti, A.R.,"An Assessment of Test Criteria in Evaluating the Performance and Integrity of Sterilizing Filters," Bulletin Parenteral Drug Assoc., Vol. 31, 1977, pp. 187-194.

(11) Treybal, R.,Mass-Transfer Operations, McGraw-Hill, New York, 1980, pp. 29-31

Peter M. Wolber, Mark A. McAllister

BUBBLE POINT DETERMINATIONS: A COMPARISON OF MANUAL VS AUTOMATED METHODS

REFERENCE: Wolber,P.M., McAllister,M.A., " Bubble Point Determinations: A Comparison of Manual vs Automated Methods" Fluid Filtration: Liquid, Volume II, ASTM STP 975, P.R. Johnston and H.G. Schroeder, Eds., American Society for Testing and Materials, Philadelphia, 1986.

ABSTRACT: With the recent development of microprocessor controlled integrity test devices, the question arises as to how existing validated parameters based on manual testing will be effected by operator independent test systems. A statistical evaluation of the bubble point results will be made using analysis of variance,randomized block design, and by interpretation of mean bubble points and standard errors. In addition, other factors including the effect of different operators and filtration system size are examined.

KEYWORDS: integrity testing, bubble point, automated integrity test devices

Microporous membrane filters are extensively used in critical applications where a defined retention characteristic is required. To insure integrity of the membrane, several different nondestructive integrity tests can be employed. The choice of the type of integrity test recommended for a membrane filter depends primarily on the manufacture's preference. This decision is often based on correlations data with the membrane's retention characteristics. The three primary nondestructive methods employed by the user and manufacturer include bubble point, diffusion, and pressure hold/pressure decay tests (1,5). All three tests can be effectively implemented to assure that the membrane is integral and that the filter is properly sealed into the filter housing, therefore avoiding process fluid bypass. Often two of the integrity tests are utilized in combination, to provide an added measure of confidence. The bubble point test is not only a method to determine filter integrity but also serves as a nominal

Mr. Peter Wolber is Technical Service Manager and Mr. Mark McAllister is a Technical Specialist at Sartorius Filters, Inc., 11 Kripes Road, East Granby, CT 06026

quantitive measure of pore size(2,4). For process facilities which utilize a wide variety of filter media and pore sizes, the bubble point offers an additional measure of safety by establishing a parameter directly related to the validated retention rating (1).

Bubble point determination for small membrane surface areas, up to 1000 cm^2, can be easily and accurately accomplished. However, when larger membrane surface areas are employed, this determination becomes increasingly more subjective (3,5). Standard methodologies for performing the bubble point test call for the gas pressure on the upstream side of the wetted filter to be slowly increased. Since the porosity of a typical membrane is approximately 70%, the wetted membrane becomes essentially a thin film of water, which allows gas diffusion directly proportional to the applied differential pressure (6). The point at which the gas diffusion ceases to be proportional to the increase in differential pressure is defined as the transition pressure phase or more commonly referred to as the bubble point pressure (3,6,7) (See Fig. 1). However, the differentiation between higher proportional gas flow and transitional gas flow is often difficult to determine visually. This results in differences between the observed and intrinsic bubble point(3). As the membrane surface area increases, this discrepancy becomes further magnified.

In order to minimize the apparent subjectivity associated with the existing bubble point test methods, automated electromechanical integrity test units were introduced in the early 1980's. Using sensitive pressure transducers to measure the upstream pressures, these integrity test devices are able to increase the level of precision of the bubble point measurement (6,8,9,10). The typical mode of operation for these devices utilizes the release of discrete quantities of gas via a solenoid valve, resulting in incremental pressure increases. The pressure measurement transferred to a logic, evaluates whether the upstream pressure increases, remains constant or decreases. If viscous gas flow is occuring, the change in the slope of pressure increases versus time is sensed, and the bubble point is registered. Although this technology offered a significant advancement in the precision of the test, the limited logic of these instruments sometimes resulted in a lack of accuracy. As described by Olson et al (6), values obtained manually were often found to be lower than bubble points obtained by the instrumental methods.

In the mid 1980's, a second generation of integrity test devices was introduced. By using a microprocessor to control the pneumatic components of the test device, a new level of sophistication has been achieved. As well a utilizing upgraded pneumatics, this new generation of devices also incorporate a sensitive pressure transducer and flow metering valve to comprehensively monitor filter response under transient conditions. This combination offers a significant advantage. It provides a means to produce sharply defined increments of pressure and simultaneously measure the volumes of gas required to maintain the pressure increase. These integrity-test systems typically allow for a short time of a few seconds between each increment, to determine the filter's response to each pressure increase. The microprocessor continuously evaluates the diffusion and pressure decay to pinpoint, most accurately, the pressure at which the gas flow begins to deviate from linearity (See figure 1). This will be the pressure at which the largest pores of the membrane are voided of fluid, and viscous gas flow and diffusion are occurring. At this point the bubble point pressure is recorded. The devices, however, will continue to increase the pressure as described to confirm that the intrinsic bubble point had been determined.

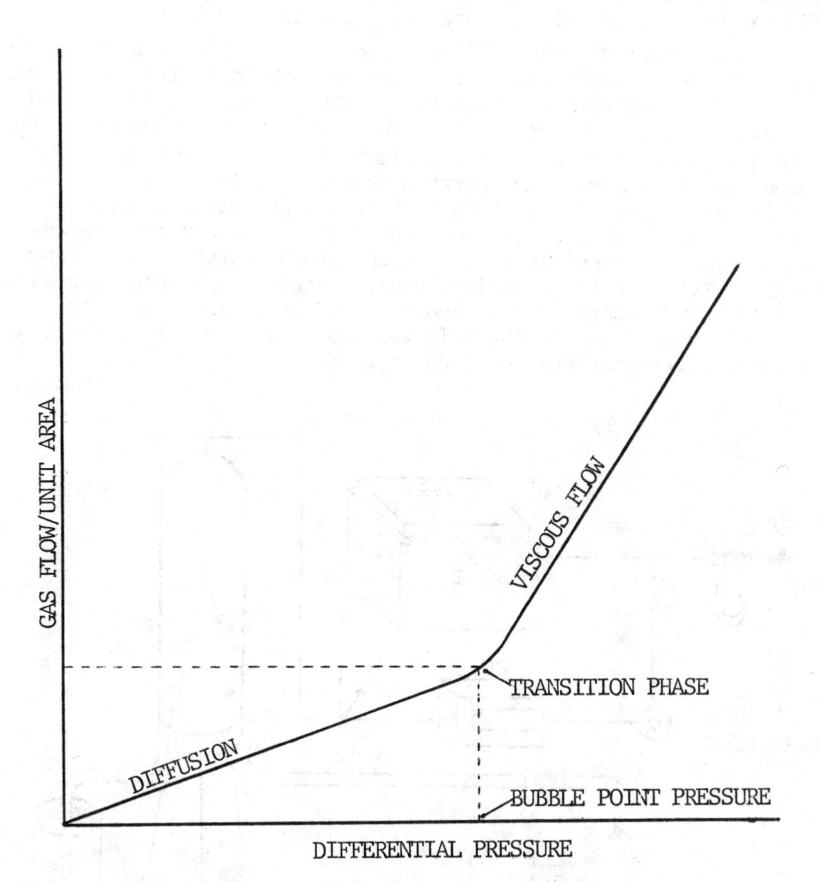

FIG. 1--Gas Flow through a fully wetted membrane with respect to applied differential pressure.

With the implementation of the new generation of microprocessor controlled integrity test devices, questions arise as to how this introduction will affect existing validation of filters and their specific product bubble points. To test the effect of automation of the integrity test, a study was conducted to compare bubble point values obtained by manual and instrumental methods. An outline of the experimental set up is drawn in Fig. 2. It incorporates a single round cartridge housing, which can be tested by both methods. For the manual method, the gas was regulated and the upstream pressure monitored by an NBS traceable gauge, with a certified accuracy of $\pm$0.01 psi.

The filter was flushed with 10 liters of deionized water and the housing and hoses were completely drained before each integrity test. The bubble point was determined by observing bubble formation from a 18mm ID hose submerged in a water filled reservoir. By switching valve V_2, the automated integrity test unit, Sartocheck II, was employed. The instrument was programmed and operated as outlined in the operational guide (11) using integrity test specifications for Sartobran-PH cartridges, which were used in the study (12). In order to generate data of statistical significance, five operators tested five different cartridges five times in a completely blind study. Each operator was fully trained and experienced in bubble point determinations. At periodic intervals the Sartocheck II was employed to generate the five bubble point determinations for each cartridge.

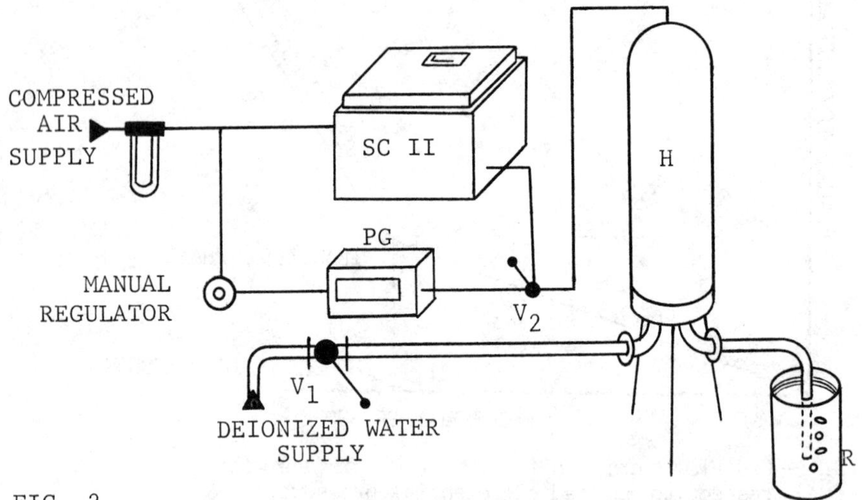

FIG. 2--
PG: NBS, Digital pressure gauge $\pm$ 0.01 psi
SC II: Sartorius Sartocheck II automated integrity test unit
V_1: Deionized water control valve
V_2: Switching valve for implementing manual or automate method
H: Filter cartridge housing
R: Water filled reservoir

The study was designed to allow interpretation of the results to analyze the bubble point determinations with respect to the following:

1. Bubble point variations by operator.
2. Bubble point variations between different operators.
3. Bubble point variations by the automated test device.
4. Differences in bubble point values between manual and automated method.

Each parameter is important in the final determinations as to whether the instrumental can be employed without invalidating existing manual method results .

TABLE 1--Determination of mean bubble points and standard errors obtained by Sartocheck II and five operators using five different single cartridge systems.*

	Cartridge System				
Operator	1	2	3	4	5
1	48.5±0.8	49.4±2.9	47.2±1.1	50.5±2.4	48.9±1.6
2	50.2±3.1	50.9±1.8	49.5±2.3	52.6±1.6	50.1±0.7
3	49.9±2.2	51.2±0.7	49.3±1.9	52.0±1.2	50.5±2.5
4	47.3±1.3	48.9±1.6	48.6±1.5	49.6±1.0	49.8±1.3
5	50.5±0.8	50.3±1.7	49.8±1.3	51.3±0.6	50.6±1.2
Sarto-check II	48.3±0.7	48.8±0.6	49.1±0.3	49.4±0.9	50.1±0.6

* Sartocheck II values were obtained in bar and converted to psi. (1 bar = 100 kPa = 14.5 psi)

The bubble point values obtained by each operator and by the instrument for the five cartridges were compiled and the means and standard errors are presented in table 1. At a glance no apparent difference or pattern can be observed. The mean bubble points were averaged and the standard errors determined comparing the manual and automated methods. No significant differences between mean bubble points can be observed, however, significant differences in the magnitude of the standard error are evident. (Table 2)

TABLE 2--Summary of the mean bubble points and standard errors calculated from values in TABLE 1, comparing manual and automated method.

Cartridge Sys.	Manual Avg Mean BP(psi)	$s_{\bar{x}}$	Automated Mean BP(psi)	$s_{\bar{x}}$
1	49.2	1.6	48.3	0.3
2	50.1	1.7	48.8	0.2
3	48.9	1.6	49.1	0.4
4	51.2	1.4	49.5	0.5
5	50.0	1.5	50.1	0.4

The standard error serves as a measure of repeatability and therefore can be considered an indicator of the precision of the methods and results. To better illustrate the variation of bubble point values obtained for the same membrane filter by different operators, the bubble point averages are presented in Table 3. The determination by manual method has 3-4 fold larger variation as compared to the automated method. Therefore, it can be stated that the automated method significantly increased the precision of the bubble point test by eliminating the operator subjectivity.

TABLE 3--Presentation of bubble point ranges using the mean and standard error calculations from Table 2

	Manual		Automated	
Cartridge	BP Range (psi)	Difference	BP Range (psi)	Difference
1	46.0-53.3	7.3	47.6-49.0	1.4
2	46.5-52.7	6.2	48.2-49.4	1.2
3	46.1-51.8	5.7	48.8-49.4	0.6
4	47.9-54.2	6.3	48.5-50.3	1.8
5	48.0-53.0	5.0	49.5-50.7	1.2

In order to statistically analyze differences between the mean bubble points, the analysis of variance using the completely randomized experimental design was employed. This test allows the analysis of the bubble point values independent of the differences in cartridges and provides a good methodology for analyzing the effect of different operators on the mean bubble point. The test results showed a statistical difference for $P<0.01$ (13).

The question now arises whether this mathematical difference will affect the integrity test results at the users end. Using a histogram presentation, the automated bubble point ranges were superimposed on the ranges obtained by the manual method (See Fig. 3). In every case the range of bubble point values obtained by the instrumental method fall into the center of the range of values obtained manually. This graphic presentation shows that the automated integrity test device not only increased the precision by reducing the variance, but also the accuracy of the measurement by providing bubble point values which were centered within the range of those manually obtained .

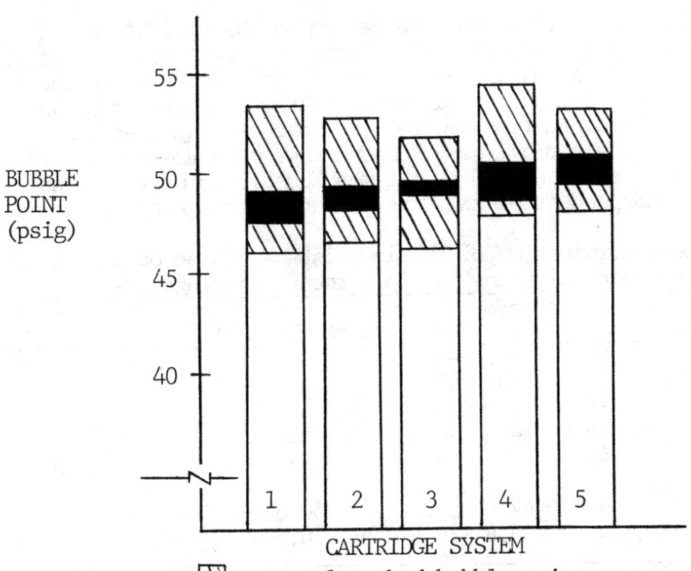

Manual method bubble point range

Automated method bubble point range (Sartocheck II values were obtained in bar and converted to psig) (1 bar = 100 kPa = 14.5 psi)

FIG. 3--Histogram presentation of bubble point ranges obtained from the single cartridge systems.

The first part of this study clearly demonstrates the value of an automated integrity test device to differentiate the diffusional properties of membrane cartridges from the true bubble points. As the membrane filtration area increases, as is the case for multiple cartridge systems, the allowable gas diffusion will increase proportionally. The operators can no longer rely exclusively on the continuous formation of bubbles to determine the bubble point value, but must differentiate between the acceptable gas diffusional parameter for the cartridges and the onset of the transitional phase.

To test the efficiency of manual and automated bubble point determinations for multiple cartridge systems, a 3 round, 3 high housing, with a total of 9 Sartobran-PH cartridges was installed into the experimental set up. The maximum allowable gas diffusion for the cartridge set up is 135ml/min., based on 15ml/min. per 10" cartridge at 80% of minimum bubble point pressure. Therefore, the operator had to determine the true bubble point against a significant background diffusion. Three operators tested the identical cartridge system five times in a blind study to determine the variability of the bubble point values and differences between operators. The data, presented in Table 4 shows in most cases a 2-fold increase in the variability as indicated by the magnitude of the standard error values compared to data obtained from single cartridge testing (See Table 1). However, results obtained by the automated test methods provided the same repeatability as witnessed by the mean bubble point values and standard error independent of the size of the cartridge system.

TABLE 4--Determination of bubble points by three operators and Sartocheck II using multiple round cartridge system*

Test	Operator 1	Operator 2	Operator 3	Sartocheck II
1	49	54	46	51
2	53	46	49	51
3	55	49	53	51
4	44	45	46	51
5	45	45	56	49
Mean ($\bar{x}$)	49.2	47.8	50.0	50.6
Std Error ($S_{\bar{x}}$)	±4.8	±3.8	±4.4	±0.9

*Bubble point values obtained by three operators and Sartocheck II testing a multiple round cartridge system five times. Sartocheck II values were obtained in bar and converted to psi. (1bar = 100kPa = 14.5psi)

A statistical evaluation using analysis of variance to determine differences in bubble point values between operators proved to be inconclusive due to magnitude of variation of the bubble point values by each operator. Differences in the bubble point ranges can be observed, in particular for operator 2, whose bubble point was consistently lower when compared to the other operators and the instrument. Despite these inconsistencies, the superimposed bubble point ranges obtained by the automated method still fall within the bubble point ranges obtained by the manual operation.

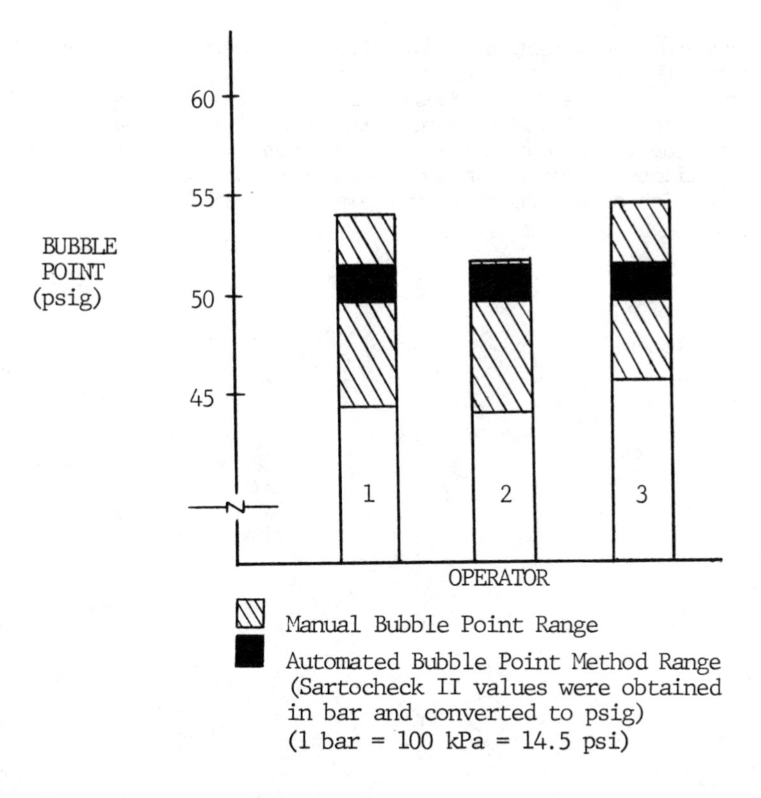

FIG. 4--Histogram presentation of bubble point ranges obtained with the large volume system.

CONCLUSION

The presented data has generated from tests conducted using 0.2 um filter cartridges which incorporate a double heterogenous membrane configuration. These were tested according to published specifications. The subjectivity, when employing manual methods to determine true bubble point values, also occurs with other filters, regardless of pore sizes, wetting fluids, membrane material, asymmetry or configurations (3,6,8).

Due to the intricate nature of the bubble point test, and the significant discrepancies obtained when employing the manual methods, careful consideration should be given to the implementation of an operator-independent test method.

The automated integrity test device provides more accurate and precise bubble point determinations as compared to those obtained when using standard manual test methods. When employing an operator-dependent method, the level of confidence further decreases as system size increases. The microprocessor-controlled device precisely and accurately performs the bubble point test, regardless of system size. As shown, the values obtained by the instrumental method also fall within the limits of those obtained manually; and thus will enable the introduction of an automated instrument with minimal required validation.

REFERENCES

(1) Karbachsch, K., " Test Methods for Integrated Filtration Systems", Technical notes from the Department of Application Technology, Sartorius GmbH, 1983.
(2) Brock, T.D., "Membrane Filtration: A User's Guide and Reference Manual", Science Tech, Inc., 1983.
(3) Hoffman, F., "Integrity Testing of Microfiltration Membranes", J. Parental Science and Technology, Vol. 38, No. 4, 1984.
(4) "Standard Method of Test for Pore Size Characteristics of Membrane Filters for Use With Aerospace Fluids, ASTM Designation F316-70", American Society Testing and Materials, Philadelphia, 1970.
(5) Trasen, B., "Non-destructive Tests for Bacterial Retentive Filters", J. Parental Drug Assoc. 33,273, 1979.
(6) Olson, W.P., Martinez, E.D. and Kern, C.R., "Diffusion and Bubble Point Testing of Microporous Cartridge Filters: Preliminary Results at Production Facilities, J. Parenteral Science Technology 35, 215, 1981.
(7) Schroeder, H.G. and Deluca, P.P., "Theoretical Aspects of Sterile Filtration and Integrity Testing", Pharmaceutical Technology 4, 80, 1980.
(8) Olson, W.P., Gatlin, L.A. and Kern, C.R., "Diffusion and Bubble Point Testing of Microporous Cartridge Filters: Electromechanical Methods", Parental Science and Technology Vol. 37, No. 4, 117, 1983.
(9) Olson, W.P., "A System for Integrity Testing of Disc and Cartridge Membrane Filters", Pharmaceutical Technology, May 1982.
(10) "Operating Instructional Manual for Sartocheck II", Sartorius Filters, Inc., Hayward, CA 94544, 1985.
(11) Eskilson, E., "Validation of Automatic Filter Integrity Test Systems, Parenteral Drug Association Poster Session, Nov. 1985
(12) "Operating Instructions for Sartorius Sartobran and Sartoclean Cartridge Membrane Filters", Technical Report #2A, Sartorius Filters, Inc., 1984
(13) Zar, T.H., "Biostatistical Analysis", Prentice Hall, Inc. 1984.

Richard V. Levy

THE EFFECT OF pH, VISCOSITY, AND ADDITIVES ON THE BACTERIAL RETENTION OF MEMBRANE FILTERS CHALLENGED WITH PSEUDOMONAS DIMINUTA

REFERENCE: Levy, Richard V., "The Effect of pH, Viscosity, and Additives on the Bacterial Retention of Membrane Filters Challenged with Pseudomonas diminuta," Fluid Filtration: Liquid Volume II, ASTM STP 975, and P.R. Johnston and H.G. Schroeder, Eds., American Society for Testing and Materials, Philadelphia, 1986

ABSTRACT: Membrane filters used in sterile production must be validated as sterilizing. Sterilizing filters, however, are used to produce solutions varying widely in chemical and physical properties, thus complicating their validation. The objective of this study was to relate bacterial retention performance of 0.22um polyvinylidene difluoride membrane (Durapore(R), Millipore Corp.) to commonly occurring solution variables and to begin to examine methods for improving the accuracy and applicability of bacterial retention testing. Experiments were undertaken to challenge 293mm filter discs with Pseudomonas diminuta suspended in a variety of diluents. The diluents, which differed with respect to pH, viscosity (glycerol, propylene glycol) and additives (Triton X-100, $MgCl_2$, $AlCl_3$) were evaluated with respect to bacterial retention. Membranes were shown to be effective in sterilizing these fluids at 30 and 50psi. Bacterial challenge tests should simulate "worst case" filter challenges using diluents which are similar to pharmaceutical solutions. They should also utilize media which best support the growth of the best bacteria and the majority of other bacteria encountered in environment monitoring.

KEYWORDS: retention testing bacteria, membranes, filters, pH, viscosity, additives

Dr. Levy is Consulting Scientist in Life Sciences Applications, Research and Development Department, Millipore Corporation, Ashby Road, Bedford, MA 01730.

BACTERIAL RETENTION OF POLYVINYLIDENE DIFLUORIDE

Introduction

Sterile filtration is used to render parenteral products bacteria free. These filters must be validated as sterilizing by conducting appropriate challenge tests. Sterilizing grade filters should perform reliably when challenged with solutions which vary widely in chemical and physical properties. Furthermore, the challenge testing should reproduce actual process conditions as much as possible.

Materials and Methods

Pseudomonas diminuta (ATCC 19146) was grown in Soybean-Casein Digest Broth (Trypticase Soy Broth - TSB) for 24 hours at 30°C without shaking in a 100ml milk dilution bottle. From this culture, 4mls were removed aseptically and inoculated into 2L of Saline Lactose Broth (SLB). The resultant culture was incubated for 24 hours at 30°C without shaking. After the incubation period, samples were taken for dilution plate quantification, purity check plating, and gram stain.

One liter of sterile 0.1% peptone water was added aseptically to a sterile reservoir vessel upstream of the test 293mm filter housing. The reservoir vessel was pressurized with filter sterilized air and sealed. The peptone culture was released quickly into the test filter housing and forced through the 0.22um polyvinylidene difluoride (PVDF) (Durapore(R)) membrane. The filtrate was collected in a sterile stainless steel collection vessel downstream. The filtrate was forced by positive, upstream pressure through a 0.45um filter. All filters were placed on trypticase soy agar plates and incubated for 48 hours. Plated filters were examined for growth of *P. diminuta* or other bacteria and enumerated. This filter was designated "Sterility Check."

Additives were prepared, filter or steam sterilized, and added aseptically to SLB cultures. Samples were taken for dilution plating after two minutes of rapid mixing. The resultant culture was aseptically transferred to the upstream stainless steel reservoir for pressurization. After the appropriate pressure was produced in the reservoir housing, either 207 KPa (30psi) or 345 KPa (50psi), the test culture was released rapidly into the 293mm flat stock filter housing filter below and forced through the sterile, 0.22um PVDF filter. The filtrate was collected in a stainless steel collection vessel beneath the filter housing. This filtrate was collected and cultured, as above, on a 0.45um filter and designated "Challenge."

One liter of sterile 0.1% peptone was passed through the test apparatus to further challenge the filter and clean the downstream housing. This filtrate was collected, as above, and designated "Chase."

Positive controls were performed to insure that the test bacterium could be recovered downstream. Bacterial suspensions were filtered through 0.45um PVDF membranes and passage was enumerated downstream. Under all conditions of pH, additives and viscosity, *P. diminuta* was recovered downstream. Both TSA and heterotrophic plate count (HPC) agars were shown to be suitable for recovery of *P. diminuta* on assay filters. R2A, which is excellent for upstream monitoring by agar plate method, was unsatisfactory for filter assays. This is probably due to the lower concentrations of nutrients which diffuse upwards into the assay filter.

Any *P. diminuta* recovered on the challenge and/or chase filters was recorded as passage. All experiments were repeated at a minimum in triplicate.

Results

The results of bacterial retention testing experiments are summarized in Table 1. Under all of these conditions, PVDF 0.22um filter was shown to be sterilizing. Increasing the viscosity by adding glycerol or propylene glycol did not affect the filter performance. The presence of divalent and trivalent ions did not adversely affect filter performance; increasing the pressure of filtration (under the conditions listed) to 50psi did not appear to affect filter performance. In the experiments reported here, all "Sterility" and "Chase" filters were negative for the presence of *P. diminuta.*

Conclusions

Bacterial retention testing was completed under the conditions listed in Table 1. PVDF 0.22um filters proved to be sterilizing under all of the conditions tested. Challenge concentrations were all under 1.0 x 10^8 cell per cm^2 effective filtration area, thus plugging of the filter by excess bacteria probably did not contribute to the performance of the filter. Log reduction values of at least 9 indicate that this filter will be sterilizing under conditions commonly encountered in sterile processing.

TABLE 1--The effects of pH, viscosity and additives on the retention of *Pseudomonas diminuta* by 0.22μm Durapore® filter.

Additive	Pressure Applied (PSI)	Bubble Point (PSI)	TOTAL CHALLENGE	LRV (Greater than)
None	30	51.5	1.9×10^{10}	10.3
None	30	52.0	1.9×10^{10}	10.3
None	30	51.5	1.9×10^{10}	10.3
None	30	51.0	1.9×10^{10}	10.3
None, pH 6.3	30	49.5	4.1×10^{10}	10.6
None, pH 6.3	30	49.5	4.1×10^{10}	10.6
None, pH 6.3	30	50.0	4.6×10^{10}	10.7
None, pH 6.3	30	51.0	4.6×10^{10}	10.7
None, pH 6.3	50	51.0	5.0×10^{10}	10.7
None, pH 6.3	50	50.5	5.0×10^{10}	10.7
None, pH 6.3	50	51.0	4.4×10^{10}	10.6
None, pH 6.3	50	51.0	4.4×10^{10}	10.6
Tris 0.05M, pH 7.4	50	51.0	1.1×10^{10}	10.0
Tris 0.05M, pH 7.4	50	52.0	1.1×10^{10}	10.0
Tris 0.05M, pH 7.4	50	52.5	7.8×10^{9}	9.9
Tris 0.05M, pH 7.4	50	51.0	7.8×10^{9}	9.9
Tris 0.05M, pH 8.0	50	50.2	3.4×10^{10}	10.5
Tris 0.05M, pH 8.0	50	50.0	3.4×10^{10}	10.5
Tris 0.05M, pH 8.0	50	49.5	6.8×10^{9}	9.8
Tris 0.05M, pH 8.0	50	49.5	6.8×10^{9}	9.8
Triton .001%	30	50.8	4.5×10^{10}	10.7
Triton .001%	30	51.0	4.5×10^{10}	10.7
Triton .001%	30	52.0	3.5×10^{10}	10.5
Triton .001%	30	52.2	3.5×10^{10}	10.5
Triton .01%	30	52.2	6.9×10^{10}	10.8
Triton .01%	30	52.6	6.9×10^{10}	10.8
Triton .01%	30	52.5	4.0×10^{10}	10.6
Triton .01%	30	52.0	4.0×10^{10}	10.6
Triton .01%	50	52.5	4.3×10^{10}	10.6
Triton .01%	50	51.0	4.3×10^{10}	10.6
Triton .01%	50	51.5	4.3×10^{10}	10.6
$MgCl_2$ 0.01M	30	52.0	2.6×10^{10}	10.4
$MgCl_2$ 0.01M	30	52.0	2.6×10^{10}	10.4
$MgCl_2$ 0.01M	30	51.5	3.0×10^{10}	10.5
$MgCl_2$ 0.01M	30	53.0	3.0×10^{10}	10.5
$MgCl_2$ 0.01M	50	50.5	5.9×10^{10}	10.8
$MgCl_2$ 0.01M	50	51.0	5.9×10^{10}	10.8
$MgCl_2$ 0.01M	50	50.8	4.8×10^{9}	9.7
$MgCl_2$ 0.01M	50	47.0	4.8×10^{9}	9.7
$AlCl_3$ 0.01M	50	50.5	3.6×10^{9}	9.6
$AlCl_3$ 0.01M	50	50.8	3.6×10^{9}	9.6
$AlCl_3$ 0.01M	50	54.0	5.4×10^{9}	9.7
$AlCl_3$ 0.01M	50	50.8	5.4×10^{9}	9.7
$AlCl_3$ 0.01M	50	49.0	1.9×10^{10}	10.3
$AlCl_3$ 0.01M	50	48.0	1.9×10^{10}	10.3
$AlCl_3$ 0.01M	50	48.0	1.1×10^{10}	10.0
$AlCl_3$ 0.01M	50	47.8	1.1×10^{10}	10.0
Glycerol, 1%	50	47.5	8.4×10^{9}	9.9
Glycerol, 1%	50	49.5	8.4×10^{9}	9.9
Glycerol, 1%	50	52.0	7.2×10^{9}	9.9
Glycerol, 1%	50	49.5	7.2×10^{9}	9.9
Glycerol, 5%	50	49.0	9.8×10^{9}	10.0
Glycerol, 5%	50	50.0	9.8×10^{9}	10.0
Glycerol, 5%	50	50.6	9.0×10^{9}	10.0
Glycerol, 5%	50	52.0	9.0×10^{9}	10.0
Glycerol, 10%	50	49.0	1.8×10^{10}	10.3
Glycerol, 10%	50	49.0	1.6×10^{10}	10.2
Glycerol, 10%	50	49.0	1.4×10^{10}	10.1
Glycerol, 10%	50	49.5	1.4×10^{10}	10.1
Glycerol, 10%	50	49.0	1.4×10^{10}	10.1
Glycerol, 20%	50	49.5	1.0×10^{10}	10.0
Glycerol, 20%	50	49.5	9.6×10^{9}	10.0
Glycerol, 20%	50	49.5	9.6×10^{9}	10.0
Propylene Glycol, 5%	50	51.0	7.3×10^{9}	9.9
Propylene Glycol, 5%	50	52.4	7.3×10^{9}	9.9
Propylene Glycol, 5%	50	51.5	5.5×10^{9}	9.7
Propylene Glycol, 5%	50	51.2	5.5×10^{9}	9.7

TANGENTIAL FLOW FILTRATION PROCESSING (CELL CONCENTRATION AND WASHING) OF RETENTION TEST ORGANISMS

Introduction

The Food and Drug Administration's proposed guidelines on sterile drug products produced by aseptic processing recommend that bacterial retention testing be conducted under conditions which "simulate actual and preferably 'worst case' conditions established as quality limits"[1]. Therefore, not only should the challenge concentration be high enough to challenge all the pores, but the bacteria should be as small as possible. The phenomenon of fragmentation and dwarfing are reproduced by growing P. diminuta in TSB for 24 hours, followed by transfer to the bacteria to SLB. SLB is a low nutrient medium which causes the bacteria to grow smaller in size and more rounded[2]. However, the bacteria remain in this medium during the sterility test. We developed a procedural modification of the classic method of sterility testing which allows P. diminuta and potentially other bacteria to be washed free of SLB and resuspended in test diluents. Our procedure was tested to determine if a tangential flow filtration system (Minitan™ System, Millipore) could be used to concentrate, wash and resuspend bacteria in selected test diluents to achieve more rigorous challenge testing.

Materials and Methods

P. diminuta was grown in TSB for 24 hours at 30°C without shaking. Two liters of SLB were inoculated with 4mLs of this culture and mixed vigorously. The SLB culture was incubated for 24 hours at 30°C. The concentration of Pseudomonas was quantified by standard dilution tube methods using both TSA and R2A agar plate counts as recovery media.

Using an autoclaved, stainless steel Minitan™ tangential flow filtration device, the saline lactose culture was concentrated by filtering the broth culture through two, stacked 0.22um PVDF plates. The filtrate was discarded while the retentate was recirculated back into the culture flask. When approximately 200mLs of the concentrated culture remained, the filtration was stopped and cell washing begun. Two liters of 0.9% saline solution (the test diluent in this case) was used to clean the cells continuously. The retentate was recirculated and the filtrate collected for disposal and sterility check.

Subsequently, the concentrated bacterial suspension was brought back up to the original volume by recirculating

1800 mLs of sterile 0.9% saline solution through the Minitan™ and into the original four-liter culture flask. This process cleans the membranes of residual Pseudomonas, while resuspending the washed bacteria in the test diluent. The concentration of washed cells was quantified, as above.

Results

We compared the recovery of *P. diminuta* grown in SLB before and after filtration. Two recovery media were tested. Table 2 summarizes the results of three replicate experiments.

TABLE 2--*Pseudomonas diminuta* recovered from saline lactose broth before diafiltration and from 0.9% NaCl retentate after diafiltration. All concentrations are the means of three replicate experiments.

	$\bar{X}$ number of *P. diminuta* recovered	
	Trypticase Soy Agar	R2A agar
	(cfu per milliliter)*	
Before Diafiltration	1.0×10^7	1.6×10^7
After Diafiltration	3.0×10^6	3.0×10^6
Percent Recovery	27	24

*cfu = colony forming units

Using the above procedure, recovery of Pseudomonas was at least 27% and 24% of the original saline lactose concentration when recovered on TSA and R2A agars. The decrease in numbers of cells present was expected since *P. diminuta* cells grown in saline lactose broth are very small and will effectively challenge the Minitan™ filters in spite of tangential forces which tend to keep them in suspension. However, in all cases the Minitan™ was able to wash out the saline lactose medium without allowing cells to pass into the filtrate. Filtrates were found to be sterile in each of the three replicate experiments. No attempt was made to test for the presence of components of the saline lactose broth in the retentate or filtrate since the PVDF membranes used in the Minitan™ were known to pass these components into the filtrate. More recent data indicate that recoveries may be improved approximately 50% by replacing 0.22um rated PVDF by 0.1um rated PVDF. We are also evaluating ultrafiltration membranes for these purposes.

Conclusions

This preliminary data indicate that a Minitan™ system can be used to concentrate, wash and resuspend bacteria in challenge test diluents. Using a more refined procedure (to increase recovery of cells) and scaled-up filtration, P. diminuta could be prepared in higher concentrations than normally available in saline lactose preparations. This method does not require tedious centrifugation for cell concentration, aseptic transfers for cell washing or cell paste preparation. The processing of two liters required approximately 40 minutes, including integrity testing of the Minitan™. Scaled-up processing to 10 liters or more would not require appreciably more time since the Minitan™ System can be expanded to filter larger volumes. The use of challenge suspensions of washed bacteria and test diluents would more closely approximate "worst" case aseptic processing conditions and simulate filtration of pharmaceutical products. Bacterial retention done in this manner should challenge the test filter more rigorously, thus lending added confidence to sterility test results.

IMPROVED RECOVERY OF *P. DIMINUTA* FROM SALINE LACTOSE BROTH CULTURES

Introduction

According to Food and Drug Administration proposed guidelines on sterile drug products produced by aseptic processing, the most important aspect of bacterial culturing related to sterility testing is the ability to promote microbial growth. In selecting such growth medium, consideration should be given to the ability of the medium to grow specific types of bacteria which have been identified during environmental monitoring and/or associated with any positive sterility test results. These bacteria are characteristically heterotrophic, gram negative bacilli, which have survived in spite of less than favorable environmental conditions.

It is generally recognized that such bacteria are recovered and cultured more effectively on low nutrient media, developed for these purposes. Examples of such media include Standard Plate Count (SPC) and Heterophic Plate Count (HPC) media[3]. More recently, Reasoner and Geldreich developed R2A medium[4] (see Table 3) which appears to be even more effective in recovering stressed organisms. Since trypticase soy medium TSA is a high nutrient, complex medium, we compared the recovery of *P. diminuta* on TSA and R2A agars.

TABLE 3--Formulae of recovery media used in these studies.

Bacto Tryptic Soy Broth
Soybean-Casein Digest Medium, USP

Ingredients per liter	
17g	Bacto Tryptone Pancreatic Digest of Casein
3g	Bacto Soytone Papain Digest of Soybean Meal
2.5g	Bacto Dextrose
5g	Sodium Chloride
2.5g	Dipotassium Phosphate
15g	Agar

Final pH 7.3 ± 0.2 at 25°C.

R2A Agar (Reasoner and Geldreich)

Ingredients per liter	
0.5g	Yeast extract
0.5g	Proteose peptone No.3 or polypeptone
0.5g	Casamino acids
0.5g	Glucose
0.5g	Soluble starch
0.05g	Dipotassium hydrogen phosphate K_2HPO_4
0.05g	Magnesium sulfate heptahydrate, $MgSO_4.7H_2O$
0.3g	Sodium pyruvate
15.0g	Agar

Adjust pH to 7.2 with solid K_2HPO_4 or KH_2PO_4 before adding agar.

Materials and Methods

Saline lactose cultures of *P. diminuta* were cultured as described in the section entitled "Bacterial Retention Studies". When present, additives were introduced into SLB cultures as described previously. TSA and R2A agar plates were prepared according to the manufacturer's (Difco, Detroit, MI) recommendations. At least four spread plates were made, each receiving 0.1ml of a 10^{-4} dilution of the SLB culture. Plates were incubated for 48 hours at 30°C and enumerated. The recovery of *P. diminuta* was determined by counting the number of colonies on each spread plate and calculating the mean number of colonies recovered on each agar. Results were expressed as the number of colony-forming units recovered on each medium per milliliter of SLB culture.

Results

The results of these experiments are summarized in Table 4. Whether additives were present or not, R2A agar recovered between 1.4 and 4.0 times as many cells as trypticase soy agar. Within experiments, increased recovery was more predictable; R2A recovered at least 20% more *P. diminuta* cells than TSA from SLB cultures.

TABLE 4--Numbers of P. diminuta recovered from saline lactose broth cultures with and without additives using spread plate recovery methods on TSA and R2A low nutrient medium. Concentrations of bacteria cells recovered are the means of at least three dilution plate counts (30-300 cells per plate).

Additive	Number recovered per ml on TSA	Number recovered per ml on R2A	Increase in recovery
None	1.5×10^7	2.2×10^7	1.5 x's
None	4.1×10^6	7.1×10^6	1.7 x's
None	4.3×10^6	8.9×10^6	2.1 x's
None	2.7×10^6	8.2×10^6	3.0 x's
None	4.4×10^6	1.5×10^7	3.4 x's
None	3.5×10^6	1.4×10^7	4.0 x's
Tris, pH 8	1.6×10^7	2.2×10^7	1.4 x's
Tris, pH 8	3.4×10^6	8.0×10^6	2.3 x's
Tris, pH 7.4	5.6×10^6	8.7×10^6	1.6 x's
Tris, pH 7.4	3.9×10^6	7.0×10^6	1.8 x's
Glycerol, 1%	4.2×10^6	1.4×10^7	3.3 x's
Glycerol, 1%	3.6×10^6	1.4×10^7	3.8 x's
Glycerol, 5%	5.0×10^6	8.3×10^6	1.7 x's
Glycerol, 5%	4.5×10^6	9.9×10^6	2.2 x's

Conclusions

Regardless of additives, in the limited number of experiments performed to date, R2A agar consistently recovers more P. diminuta from saline lactose broth cultures than trypticase soy agar. The use of R2A or equivalent medium (e.g., HPC) to quantify the challenge inoculum should be encouraged. Trypticase soy medium underestimates the number of P. diminuta in challenge suspensions.

SUMMARY

Polyvinylidene difluoride membrane flat stock was shown to be sterilizing under all of the test conditions. Total bacterial challenge concentration was at least 1.0×10^9 cells. The presence of a nonionic detergent, divalent and trivalent ions did not affect filter performance. Increasing the viscosity or pressure of filtration did not affect performance adversely. At pH values of 6.3, 7.4 and 8.0, P. diminuta did not pass through the filters. Attempts were made to devise a method for the diafiltration of P. diminuta cells. Recovery of saline lactose grown P. diminuta cells was over 24%. This diafiltration process with increased efficiency could be used to wash bacterial cells free of growth medium, and permit the cells to be resuspended in a variety of diluents. Therefore, bacterial challenge testing could be run without the influence of media constituents and metabolic by-products. Recovery of retention test bacteria and other filtrate contaminating bacteria can be enhanced by replacing trypticase soy agar with a lower nutrient, lower ionic strength medium, R2A or HPC designed for the recovery of stressed bacteria from water environments. Improvements in sterility testing can be expected to improve the reliability and applicability of such testing.

ACKNOWLEDGMENTS

The author acknowledges with appreciation the help of Marsha Cleversey and Paul Hatch in the technical work they performed, and the help of Frances Carlson and Deborah Moschella in the preparation of this paper.

REFERENCES

1. Food and Drug Administration, "Proposed Guideline on Sterile Drug Products Produced by Aseptic Processing," January 1985.

2. Leahy, T.J. and Sullivan, M.J., "Validation of Bacterial Retention Capabilities of Membrane Filters," Pharm. Tech., Vol. 2, 1978, pp. 65-75.

3. Standard Methods for the Examination of Water and Wastewater, 16th Edition, A.P.H.A., A.W.W.A., W.P.C.F., 1985.

4. Reasoner, D.J. and Geldrich, E.E., "A New Medium for the Enumeration and Subculture of Bacteria from Potable Water," Appl. Environ. Microbiol., Vol. 49, 1985, pp. 1-7.

Wayne P. Olson and James R. Greenwood

MICROBIAL RETENTION BY DEPTH FILTERS WITH PENDANT HYDROPHOBIC LIGANDS

REFERENCE: Olson, W. P. and Greenwood, J. R., "Microbial Retention by Depth Filters with Pendant Hydrophobic Ligands," Fluid Filtration: Liquid, Volume II, ASTM STP 975, P. R. Johnston and H. G. Schroeder, Eds., American Society for Testing and Materials, Philadelphia, 1986.

ABSTRACT: The bacterial and fungal holding capacity of fibrous nylon depth filters and filter aids is increased by orders of magnitude when long alkyl ligands are grafted to the matrix. Organism retention is not affected by human albumin or beer, but an oil/water emulsion (homogenized milk) interferes with the binding of cells. Viruses also are retained, but with low efficiencies for brief dwell times.

KEYWORDS: bacterial binding, beer filtration, depth filtration, hydrophobic interaction, hydrophobic ligands, ligand grafting, pendant hydrophobes, protein filtration, virus binding

THE MECHANISM OF MICROBE CAPTURE AND RETENTION BY FILTERS

Capture

The mechanics of particle capture by a filter or a filter aid are widely accepted as inertial impaction, sieving, and "diffusion." Inertial impaction occurs when a particle or large macromolecule cannot follow the fluid streamlines into a pore, and strikes the filter matrix. Particles larger than a pore are said to be captured by a sieving mechanism; a related event is "bridging," whereby particles smaller than a pore aggregate to occlude a pore. "Diffusion" refers to the Brownian or other movements of particles (usually much smaller than the pore) toward the pore wall

Mr. Olson is Senior Associate in Technical Services at Hyland Therapeutics Division of Travenol Laboratories, 4501 Colorado Blvd., Los Angeles, CA 90039. Dr. Greenwood is Director, Public Health Laboratory, 1729 West 17th Street, Santa Ana, CA 92706.

or fiber/filter aid surface. The mechanisms of capture do not account for the retention of particles, for example, particle bounce. And why are particles washed away, or not? Sieving and bridging serve to explain the retention of some particles, but tell us nothing about the retention of small particles, especially bacteria, fungi, and viruses.

Retention

Ionic: The attachment of *Vibrio alginolyticus* to hydroxyapatite veries with salt molarity and valence of the cation (1). The data of Gordon and Millero indicate divalent cations, such as Mg(II), form a bridge between negatively charged vibrios and the negatively charged hydroxyapatite. Maximal binding occurs at cation concentrations sufficient to coat the bacteria or the solid surface, whichever has the higher affinity for the cation, and at higher concentrations of the cation, some repulsion occurs as both the bacteria and the hydroxyapatite acquire a positive charge. As one would expect, the effect of divalent cations is greater than that of monovalent cations. The effect of the electrolyte might be said initially to favor aggregation or flocculation, and at higher concentrations to favor peptization. There are many examples of bacterial or viral retention on depth filters to which positively charged moieties have been coupled (2,3).

Hydrogen bonds: Poliovirus 1 and coxsackievirus B3, sorbed onto Zeta-Plus 30S, cannot be eluted successfully with acid or alkali or salts alone. Best elution of the viruses is with 4M urea made 50 mM in lysine and with the pH adjusted to 9. Urea interdicts H-bonds. Experimental details are in (3).

Hydrophobic: Neither detergents, nor salts, nor ethyl alcohol at pH 4 elute poliovirus from mixed esters of cellulose; however, $MgCl_2$ or NaCl PLUS ethanol at pH 4 or detergent cause the virus to elute (4). Consequently, both electrostatic and hydrophobic interactions are implicated in the retention of this virus by mixed esters of cellulose. Rosenberg (5), in the introduction to his work on variations in the cell surface hydrophobicity of *Serratia marcescens*, indicates that hydrophobicity of some cell surfaces accounts for the adhesion of bacteria to oil, one another, epithelial cells, teeth, nonwettable plastics and marine sediments.

Affinity: Affinity interactions are taken by the biochemist to refer to the interactions of an homologous pair, e.g., antigen-antibody, enzyme-inhibitor, enzyme-cofactor, etc. Many molecules, e.g., acetylcholine, also have specific receptors on the surface of cells. On *Escherichia coli* grown on 0.2% maltose there is a lambda receptor that binds to starch. Ferenci (6) coupled starch to agarose and found that 94% of 2.3×10^9 cells bound per mL of the starch-substituted agarose; only 1.1% of the bound cells eluted in 200 bed volumes of medium. Salt, pH and temperature had little effect on the retention of cells by the matrix. This likely is a rare mechanism for microbial retention on a filter. It would be of interest to know whether maltose, or isomaltose, or both disaccharides, when mixed with the cells prior to addition to the matrix, reduces significantly the proportion of cells binding.

Other: Irvin et al. (7) showed that *Alysiella bovis* adhered to glass, and probably to bovine epithelium, via a 16.5 kdalton glycoprotein, 17% of which was carbohydrate. The nature of the interactions was not determined.

HYDROPHOBIC INTERACTIONS

With Solid Surfaces

Organics: Penicillins and cephalosporins sorb well to nonionic styrenedivinylbenzene regardless of salt concentrations, although methanol decreases the sorption (8). As one would expect from partitioning behavior, acidic penicillins and cephalosporins sorb well at acidic pH, but not at alkaline pH when they are ionized. Similarly, the pharmaceutical literature contains evidence of the sorption of hydrophobic drugs from solution into plastic tubing.

Proteins: Many proteins sorb to hydrophobic surfaces. The styrenedivinylbenzene ion exchange matrices are notorious for the irreversible binding and denaturation of many proteins. IgG is not denatured, but sorbs via the F_C region ("tail") to polystyrene latex beads and polystyrene test tube walls; this is how many immunologic diagnostics are made. But as avidly as IgG is bound to polystyrene, it can be desorbed with 55-60% dimethylsulfoxide (DMSO)(9). In thiscase, DMSO interdicts the hydrophobic interaction, i.e. DMSO has a higher affinity for the plastic than does the protein. Tween 20 or Tween 80, added to polystyrene wells in a reaction plate, either before or after the addition of peroxidase, prevented the irreversible sorption and inactivation of the enzyme (10). The Tweens are ethylene oxide adducts, and we suspect that ethylenic groups orient against the plastic and the oxygens are distal to the plastic, facing the aqueous solution.

Cells: Rosenberg (11) has shown that bacteria with hydrophobic surfaces, e.g., *Staphylococcus aureus*, adhere to a smooth polystyrene surface, whereas *Staphylococcus albus*, the surface of which is hydrophilic, does not stick. Similarly, *Candida tropicalis*, with a more hydrophobic surface than *C. albicans*, adhered better than *C. albicans* to hydrophobic surfaces (12). Large beads of quite hydrophobic polymerized poly(vinylpyridinium halide) reduced colony-forming-units of Escherichia coli, Pseudomonas aeruginosa, Salmonella typhimurium, Staphylococcus aureus, and Streptococcus faecalis from 10^7 or 10^8 per mL by about 2 logs. As we will see, a 2 log reduction is not impressive.

With Ligands on Agarose Beads

Lipids and Proteins: Deutsch et al. (14) coupled dodecylamine to agarose via the amino group. When these particles were added to plasma, the lipoproteins were removed almost entirely by the resin. Lipoproteins are, for our purposes, oil-in-water emulsions, and we assume that the lipoprotein particles attached to the alkyl ligands coupled to the agarose. Hofstee and Otillio (15) and Hofstee (16) developed evidence of the affinity of various proteins for ligands that represent a hydrophobicity gradient. The hydrophobic ligands, unlike broad hydrophobic surfaces, tend not to denature proteins such as enzymes. Therefore, such enzymes as are not denatured by sorption to an alkyl ligand may be immobilized in that way, and made into an enzyme reactor (17).

Viruses: Aminodecyl-agarose sorbs the hepatitis virus from human plasma (18); a C_{16} to C_{18} ligand with a terminal amino group provides best results, although sorption of the virus is slow and is best done in a mixed batch mode (19). Satellite tobacco necrosis virus sorbed to dodecyl agarose in high salt, and was purified, i.e., released from the matrix, with a descending salt gradient (20). Parenthetically, we note that hydrophobic interactions are promoted by increasing levels of electrolyte, whereas ionic interaction is interdicted at high levels of electrolyte.

Cells: Baker's yeast was retained on naphthyl agarose at high salt concentrations, but eluted with a descending salt gradient (20). Some marine bacteria could be retained by octyl agarose (21). Erythrocytes also can be immobilized by interaction with alkyl ligands coupled to a solid support (22).

These examples serve to indicate how microbes may be retained on a filter or other surface. The effects of viscous drag in laminar flow of a liquid over a particle sorbed to a surface, and the effects of turbulent flow on that particle, are not considered here. Our primary interest is in the hydrophobic interactions.

FABRICATION AND EVALUATING FILTERS WITH HYDROPHOBIC LIGANDS

Fabrication

Most coupling chemistry for the grafting of alkyl ligands to a filter matrix require hydroxyl, amino, sulfhydryl, carboxyllic acid, vinyl, or other reactive groups on the surface of the filter matrix. For example, polyethylene and polypropylene must be substituted before efficient ligand coupling by wet chemistries can be done.

Cyanogen Halide: The CNBr method for activation of polysaccharides has been used by Hofstee (16) and many others. Adjacent hydroxyl groups are required for successful generation of the intermediate (22). Coupling done in the CNX manner is highly unstable at alkaline pH. The technique ought not to be used with filters intended for production use at a pH greater than 6.5.

Acid Chloride: An ester is formed with hydroxyl groups (as on cellulose). The method can be done rapidly in pyridin e, but the ester linkage tends to hydrolyze in aqueous acid or alkali. Such a matrix requires pre-use cleanup with ethanol or methanol.

Bis-Glutaraldehyde: If nylon fibers are reacted for a few minutes with 0.1 N HCl, many free $-NH_2$ groups are generated. After rinsing the fibers, the amino groups can be reacted with glutaraldehyde, washed, and then reacted with the desired alkylamine. As with the CNBr method, two steps are required.

Bis-Oxirane: We reacted 1,4-butanediol diglycidyl ether with nylon fibers, and then with octylamine, to produce the ligand shown in Figure 1. Microbial binding with this ligand is reported elsewhere in this communication.

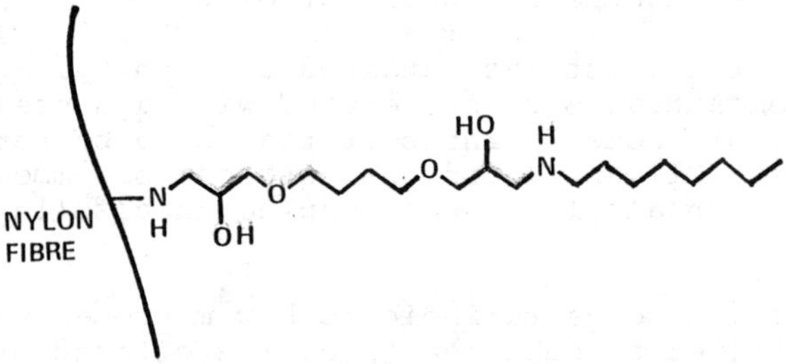

FIG. 1--Structure of a bis-oxirane-derived pendant hydrophobic ligand on nylon fibers.

Evaluation

Bacteria: Two grams of the substituted or the unsubstituted nylon was placed in a 3 mL syringe barrel and rinsed with distilled water. All rinsewaters were nontoxic for the bacteria under test. The indicated counts (Inoculum) of the test organisms (*Bacillus subtilis*, *Pseudomonas diminuta* or Staphylococcus aureus) were added to 2 mL of the indicated medium and eluted with 20 mL of the medium. The data are reported in Table 1.

Table 1--Removal of bacteria from phosphate-buffered saline (PBS), distilled water, whole milk, beer, and 5% human serum albumin (HSA). Control is filtrate from unsubstituted nylon fibers, and Test is from alkylated nylon fibers

		Bacteria per mL		
Medium	Organism	Inoculum	Control	Test
PBS	*P. diminuta*	120,000	24,000	0
	S. aureus	320,000	72,000	1
Water	*P. diminuta*	50,000	41,000	0
	S. aureus	330,000	39,000	3
Milk	*P. diminuta*	140,000	33,000	32,000
	S. aureus	480,000	67,000	24,000
Beer	*P. diminuta*	80,000	29,000	0
	S. aureus	440,000	8,000	140
HSA	*B. subtilis*	30,000	11,000	90
	P. diminuta	80,000	39,000	0
	S. aureus	280,000	69,000	2

Generally, the data indicate a 3 to 4 log improvement in bacterial removal attributable to the alkylation of the nylon fibers. The emulsified lipids in milk displaced substantial proportions of the bacteria from the test filters, likely because the milk fat had a higher affinity for the alkyl ligands than did the bacteria. We assume that the ligands do not react only with the cell surface, but extend into the double-layered phospholipid cell membrane.

Results with diatomaceous earth (DE) and C_{18}-substituted DE (Sep-Pak, Waters Assoc., Milford, Mass.) were not as good. Neither bound *Proteus* and other cells nearly as well as the nylon. Apparently, the positive charge of free (unreacted) amino groups on the nylon surface is needed to attract the cells, and the pendant hydrophobe then penetrates the cell membrane to effect firm retention.

Fungi: When Baker's yeast was added to Control or Test syringe barrels, virtually all yeast was removed by both matrices, likely because the cells are large. However, when the nylon fibers were chopped to short lengths with a razor blade, added to 20 mL of PBS in which the yeast was suspended, the tubes shaken by hand for 5 minutes, and the supernatants recovered by passing the suspensions through fine-mesh nylon netting, we obtained different results. Of the 52,000 yeast cells added to the Control, 520 remained in suspension. Only 41 remained in the supernatant of the Test eluate. In this instance, the short nylon fibers were used as filter aids and improved yeast removal about tenfold.

Viruses: A high-titered cell culture preparation of Type 1 Herpesvirus was diluted in either sterile water or albumin and poured onto the filter in the syringe barrel. The plunger then was inserted and the filtrate was expelled into test tubes. The data are given in Table 2.

Table 2--Herpesvirus Type 1 removal from solution by alkyl-substituted nylon fibers

	Virus Titer as Log_{10}	Titer Reduction as Log_{10}
Unfiltered virus in 5% albumin	5.2	1.2 ± 0.5
Unfiltered virus in distilled water	5.2	2.0 ± 0.6

These data are less encouraging than those for Hepatitus B surface antigen removal (18). We suspect that the difference in dwell time (negligible for these experiments and up to 16 hours in successful work, 18,19) accounts for the dramatic differences in efficiencies. Presumably, the probability of a small virus colliding with a fiber is far less than the probability of a bacterium (which is orders of magnitude larger than herpes or hepatitis viruses) colliding with the same fibers, when equivalent mixing is done.

In the short-term experiments, unsubstituted nylon was equally as effective as the alkylated material, which also contains amino groups. For viruses, short-term exposure to immobilized amino groups does not lead to significant retention.

CONCLUSIONS

Substitution of depth filters with long alkyl groups increases their capacity for binding bacterial and fungal cells by orders of magnitude, but in high-flux deadend filtration, viruses are not efficiently removed. Fixed positive charges, as on the surface of nylon, promote binding. Presumably, increased bacterial binding occurs because the alkyl ligands extend into the cell membrane. There is a potential for improved efficiency in the removal of spoilage organisms from wines and unpasteurized beers, as well as in the routine filtration of cosmetics and pharmaceuticals, based on a combination of ionic and hydrophobic interactions.

REFERENCES

(1) Gordon, A. S. and Millero, F. J., "Electrolyte Effects on Attachment of an Estuarine Bacterium," Applied and Environmental Microbiology, Vol. 47, No. 3, Mar. 1984, pp. 495-499.

(2) Keswick, B. H., "Survival of Enteric Viruses Adsorbed on Electropositive Filters," Applied and Environmental Microbiology, Vol. 46, No. 2, Aug. 1983, pp. 501-502.

(3) Chang, L. T., Farrah, S. R., and Bitton, G., "Positively Charged Filters for Virus Recovery from Wastewater Treatment Plant Effluents," Applied and Environmental Microbiology, Vol. 42, No. 5, Nov. 1981, pp. 921-924.

(4) Shields, P. A. and Farrah, S. R., "Influence of Salts on Electrostatic Interactions Between Poliovirus and Membrane Filters," Applied and Environmental Microbiology, Vol. 45, No. 2, Feb. 1983, pp. 526-531.

(5) Rosenberg, M. "Isolation of Pigmented and Nonpigmented Mutants of Serratia marcescens with Reduced Cell Surface Hydrophobicity," Journal of Bacteriology, Vol. 160, No. 1, Oct. 1984, pp. 480-482.

(6) Ferenci, T., "Affinity Immobilization of Escherichia coli: Catalysis by Intact and Permeable Cells Bound to Starch," Applied and Environmental Microbiology, Vol. 45, No. 2, Feb. 1983, pp. 384-388.

(7) Irvin, R. T., To, M., and Costerton, J. W., "Mechanism of Adhesion of Alysiella bovis to Glass Surfaces," Journal of Bacteriology, Vol. 160, No. 2, Nov. 1984, pp. 569-576.

(8) Salto, F. and Prieto, J. G., "Interactions of Cephalosporins and Penicillins with Nonpolar Macroporous Styrenedivinylbenzene Copolymers," Journal of Pharmaceutical Sciences, Vol. 70, No. 9, Sept. 1981, pp. 994-998.

(9) deBruin, H. G., Van Oss, C. J., and Absolom, D. R., "Desorption of Protein from Polystyrene Latex Particles," Journal of Colloid and Interface Science, Vol. 76, No. 1, July 1980, pp. 254-255.
(10) Berkowitz, D. B. and Webert, D. W., "The Inactivation of Horseradish Peroxidase by a Polystyrene Surface," Journal of Immunological Methods, Vol. 47, 1981, pp. 121-124.
(11) Rosenberg, M., "Bacterial Adherence to Polystyrene: a Replica Method of Screening for Bacterial Hydrophobicity," Applied and Environmental Microbiology, Vol. 42, No. 2, Aug. 1981, pp. 375-377.
(12) Minagi, S., Miyake, Y., Inasaki, K., Tsuru, H., and Suginaka, H., "Hydorophobic Interaction in Candida albicans and Candida tropicalis Adherence to Various Denture Base Resin Materials," Infection and Immunity, Vol. 47, Jan. 1985, pp. 11-14.
(13) Kawabata, N., Hayashi, T., and Matsumoto, T., "Removal of Bacteria from Water by Adhesion to Cross-Linked Poly(Vinylpyridinium Halide)," Applied and Environmental Microbiology, Vol. 46, No. 1, July 1983, pp. 203-210.
(14) Deutsch, D. G., Fogleman, D. J., and von Kaula, K. N. "Isolation of Lipids from Plasma by Affinity Chromatography," Biochemical and Biophysical Research Communications, Vol. 50, No. 3, 1973, pp. 758-764.
(15) Hofstee, B. H. J., and Otillio, N. F., "Non-Ionic Adsorption Chromatography of Proteins," Journal of Chromatography (Chromatographic Reviews), Vol. 159, 1978, pp. 57-69.
(16) Hofstee, B. H. J., "Selective Aromatic-Hydrophobic Binding and Fractionation of Immunoglobulins by Means of Phenyl-$(CH_2)_n$-NH-Substituted Agaroses," Biochemical and Biophysical Research Communications, Vol. 91, No. 1, 1979, pp. 312-318.
(17) Kaetsu, I., Kumakura, M., and Yoshida, M., "Immobilization of Enzymes or Bacterial Cells," U.S. Patent 4,272,617, June 1981.
(18) Neurath, A. R., Lerman, S., Chen, M., and Prince, A. M., "Hydrophobic Chromatography of Hepatitis B Surface Antigen on 1,9-Diaminononane or 1,10-Diaminodecance Linked to Agarose," Journal of General Virology, Vol. 28, 1975, pp. 251-254.
(19) Andersson, L.-O., Borg, H. G., and Einarsson, G. M., "Method of Removal of Hepatitis Virus," U.S. Patent 4,168,300, Sept. 1979.
(20) Hjerten, S., Rosengren, J., and Påhlman, S., "Hydrophobic Interaction Chromatography. The Synthesis and the Use of Some Alkyl and Aryl Derivatives of Agarose," Journal of Chromatography, Vol. 101, No. 2, Dec. 1974, pp. 281-288.
(21) Dahlbäck, B., Hermansson, M., Kjelleberg, S., and Norkrans, B., "The Hydrophobicity of Bacteria--An Important Factor in Thir Initial Adhesion at the Air-Water Interface," Archives of Microbiology, Vol. 128,

1981, pp. 267-270.
(22) Halperin, G. and Shaltiel, S., "Homologous Series of Alkyl Agaroses Discriminate Between Erythrocytes from Different Sources," *Biochemical and Biophysical Research Communications*, Vol. 72, No. 4, 1976, pp. 1497-1503.

Raymond C. Lukaszewicz, Theodore H. Meltzer, Ros Davis, and Louis Bilicich.

FLUOROCARBON FILTER MATERIALS: CHARACTERISTICS, DISTINCTIONS, AND USES

REFERENCE: Lukaszewicz, R. C., Meltzer, T.H., Davis, R., and Bilicich, L, "Fluorocarbon Filter Materials: Characteristics, Distinctions, and Uses", Fluid Filtration: Liquid, Volume II, ASTM STP 975, P. R. Johnston and H.G. Schroeder, Eds., American Society for Testing and Materials, Philadelphia, 1986

ABSTRACT: Over the past several years much confusion has existed over the proper terminology used to describe Fluorocarbon and Fluoropolymer filter types and their application areas versus traditional membrane polymers such as cellulosics. Since distinct differences exist in extractable levels, hydrophobicity, chemical compatibility and ultimate performance which are of critical importance this paper defines the inherent differences between filter manufacturers' offerings, new products that are available and suggests areas in which these new generation filters should be employed effectively.

KEYWORDS: perfluorinated polymers, florinated polymers, perfluoroalkoxy polymer, fluorocarbon, fluoropolymer

INTRODUCTION

From the very inception microporous membrane manufacture utilized the rich variety offered by polymer science in fulfillment of its product needs. The early mixed esters of cellulose polymers were used for their film-forming properties. Their blending conferred upon the microporous film a necessary degree of flexibility. One ester type served as an internal plasticizer for the other; an art borrowed from the lacquer maker. The use of cellulose triacetate polymers permitted a less sensitive autoclaving because of their higher softening point, and in the absence of the strong hydrogen-bonding nitrate group common to the mixed esters of cellulose permitted the resolution of human serum in electrophoretic diagnoses. A copolymer of vinyl chloride and acrylonitrile (vinyl acrylate copolymer), a formulation designed to be resistant to the oxidative assaults inflicted on silver/zinc battery separators was devised as a membrane coating on nylon parachute cloth. The reinforcement of the nylon support permitted folding of the microporous arrangement and made possible the pleated membrane cartridge.

There followed the development of the polysulfone, polyvinylidene fluoride, and nylon microporous membranes whose properties include an ability to withstand direct in-line steaming.

Each of the polymers used offers its own assortment of properties. Each has its advantages and limitations. The mixed esters of cellulose may absorb proteins when this is undesired, but can, for precisely this reason, be used to retain viruses. The cellulose triacetate can be plasticized, as with glycerine, to be pleatable at elevated temperatures; yet this practice may entail extractables that require removal. The vinyl acrylate polymer offers commendable resistance to oxidizing sanitization agents; but as a coating may insufficiently wet the supporting cloth so that integrity testability is compromised. Polyvinylidene fluoride filters have their inherent hydrophobicity overcome by the chemical grafting of a wettable surface; still extractables are detectable. Nylon fibers often require no wetting adjuncts, possibly due to the presence of polar surface grouping, as could be created by formic acid hydrolysis, but the very hydrolysability of nylon precludes its long term use in low pH environments, and its susceptibility to oxidizing agents is marked. The polysulfone polymer, reassuringly stable and oxidation resistant, even to high concentrations of ozone, exhibits a distressing propensity to undergo catastrophic stress-cracking when permitted to dry directly after encountering alcohols. As would be expected, then, the use of any particular polymeric membrane filter must be made with both its limitations and strengths in mind. This applies, too, to the microporous fluorocarbon filters. However, their applicable range of uses when compared to other polymeric types is significantly more broad.

There are applications where extreme inertness to aggressive chemical reagents are required. Microporous membranes of polytetrafluoroethylene (PTFE) offer this extreme inertness (Such filters are usually made of Teflon®, E. I. Du Pont de Nemours Company's registered trade name for this polymer). There are applications where extreme hydrophobicity would seem to be needed. Microporous PTFE filters are indicated. Yet, even here there are slight tradeoffs in the construction of the membranes, and, however slight, in the cartridges built of them. Additionally, there is opportunity for confusion arising from the use of the generic terms fluorocarbon and fluoropolymer. A review of the situation is in order.

Perfluorinated Polymers

The fluorocarbon properties of pronounced inertness and hydrophobicity that are the qualities of interest in the application of microporous membranes are expressions of the fluorine/carbon bond. It can be ruptured by excessive heat, and by the action of molten alkali metals such as sodium. Normally, however, it is supremely inert to solvents, oxidizers, reducers, and usually aggressive chemical reagents. The true perfluorinated polymers are molecular compositions of carbon-to-carbon and fluorine-to-carbon bonds alone. They eventuate as various molecular weight polymers from the free radical vinyl polymerizations of tetrafluoroethylene monomer, $F_2C = CF_2$. The final polymeric structures contain no carbon-to-hydrogen or carbon-to-other non-fluorine substitutes on the long polymeric chains. Indeed the prefix "per" implies just that, a complete presence of fluorine atoms as substituents wherever the structural bonding creates sites for substituents to exist. The total saturation of the carbon valences (other than those comprising the carbon-to-carbon polymer backbone) by fluorine atoms, endows the resulting polyfluorocarbons with the ultimate resistivity and hydrophobicity of the perfluorinated structure.

Unfortunately, perfluorinated polymers, except as moderated somewhat by the manipulations of crystallinity and molecular weight, are so intractable as almost to defy the molding, forming, or shaping possible with other polymers by the application of heat and/or pressure. Some possibilities do exist of joining, say PTFE particles to one another as by compression molding. The temperature requirements are higher than for most thermoplastics. Also, the temperature window is very narrow. Below the required temperature, flow does not occur. Above the narrow range, the polymer decomposes. The evolving fluorine gas is savage in its attack on the surfaces it encounters; indeed, it may serve to degrade the polymer itself. Corrosion resistant materials such as Monel, Duranickel, or Hastalloy C, are preferred in the forming of the perfluorinated materials.

Fluorinated Polymers

Given the intractability of the perfluorinated polymers, it is often considered more practical to leaven the polymeric composition by use of materials not totally composed of fluorine substituents. The resulting polymers are fluorinated, but only to degrees; not totally fluorinated as implied by the prefix "per", and therefore not totally possessed of the ultimate chemical inertness and hydrophobicity of the fluorine-to-carbon bond materials. On the other hand, they are more amenable to processing.

Such polymers can be prepared from monomers as clorotrifluoroethylene $Cl(F)C{=}CF_2$, wherein a less resistant chlorine/carbon bond is substituted for one of the fluorine/carbon bonds of tetrafluoroethylene, $F_2C{=}FC_2$; this monomer is less fluorinated than tetrafluoroethylene. Its polymers are, therefore, easier to fabricate but only slightly less as hydrophobic or chemically resistant. Polymers can be prepared from vinylidene fluoride. "Vinylidene" is a term denoting two like substituents, in this case fluorine atoms, resident on the same vinyl carbon atom, (one of the two carbons constituting the double bond).* The corresponding polymers, containing only half the possible amount of fluorine atoms of the perfluorinated materials, are more easily formed but substantially less inert and hydrophobic. It is also possible to prepare polymers containing fluorine atoms from vinyl fluoride monomer, $H(F)C{=}CH_2$. In this case, only one of the four possible constituents is a fluorine atom.

*Vinylidene difluoride is, therefore, an unnecessary redundancy, technically incorrect but widely used.

Perfluoroalkoxy Polymer (PFA)

There is a variant of the perfluorinated polymer that bears a single alkoxy substituent, an alkylated ether linkage, about once for every ten possible fluorine atoms. Strictly speaking it does not merit the "per" appelation, which implies the ultimate degree of thoroughness here of fluorination. Yet, this polymer is more fluorinated than any currently available except the polytetrafluoroethylene. It has proven a most useful interfacing material as a melt in joining PTFE to PTFE, and is in wide use in PTFE cartridge construction for this reason.

Fluoropolymer Terminology

It is regrettable that the term "fluoropolymer" or even the careless usage of the Du Pont registered trade name Teflon™ has come into such inexact circulation. The result has all too often served to cloak as an "all PTFE" or "all Teflon®" cartridge, a filter device that does contain fluorinated polymeric material, but hardly in the degree of completeness or construction that the name implies.

Cartridges composed completely of PTFE except for the use of PFA and or FEP as a joining material, used in the form of a hot melt, are, however, available. These should not be confused with the lesser fluorinated polymeric compositions such as those fabricated of polyvinylidene fluoride, or those containing Teflon microporous membranes laminated to non-PTFE materials such as polyethylene or polypropylene, or others utilizing adhesions by hot melt polypropylenes [1].

All Fluorocarbon Cartridge

Until 1985 fluorocarbon membranes could be only purchased in disc form for small volume filtration applications or in cartridges for large processing volumes. The fluorocarbon filter had to be married with other polymers such as polypropylene, polyethylene, or polvinylidene fluoride in order that a seal could be made which would permit industry accepted bubble point or diffusion integrity testing.

Now there is on the market a patent pending "all Teflon®" cartridge, as described by the E.I. Du Pont Company itself. It is composed of a core, outercage, and end caps of Teflon®. The membrane is microporous Teflon® pleated in conjunction with support layers of Teflon® screens. Adherence of one part to another, as also of the membrane seaming, is of Teflon®.

Microporous PTFE Membranes

Microporous PTFE is available to filter manufacturers, as from W. L. Gore and Associates of Elkton, MD (Gore-Tex) since 1972, and one filter manufacturer purchases media from Garlock Inc. of Newtown, PA. This microporous material is usually utilized in laminated form for the better handling properties and superior dimensional stability thus provided, although it is available in the form of pure PTFE as well. The microporous membrane is laminated to polyethylene to polypropylene for somewhat better handlings qualities, and to polyester. The substrate materials are usually available in both woven web and non-woven mat forms. Indeed, even polyurethane foams are used. The microporous membrane is produced in various pore-size ratings, namely, 0.02, 0.1, 0.2, 0.45, 1.0, 3.0, 5.0, and 10-15 um (Micrometer).

The pore sizes of the submicronic PTFE membranes are characterized using the ASTM F316 bubble point method. Because of the hydrophobicity of the polymer, the wetting liquid is customarily some short-chain alcohol, usually absolute but on occasion used as an aqueous solution. Methanol, ethanol, propanol, and even 25 percent tertiary butanol may be used. Employing 95 percent ethanol, Garlock Inc., find the relationships in Table 1:

TABLE 1--Pore Size to Bubble Point Relationship

Pore-Size (um)	Minimum Bubble Point (PSI)
0.2	13
0.45	7
1.0	3

W. L. Gore and Associates also characterized their microporous Teflon® filters by means of an alcohol (methanol) bubble point. The 3.0, 5.0 and 10-15 um sizes are measured using a manometer rather than a gauge, and the bubble point values are listed in terms of inches of mercury, or in inches of water rather than in fractions of one PSI, in order to achieve accurate readings.

Pore-size rating is assigned on the basis of matching bubble point values to those given by conventional phase inversion membranes. This unjustified method of pore-size rating, by transferring the values given by one polymer to another, is by no means unusual among filter manufacturers who use polymers other than the cellulosic esters. It presupposes, from a matching of filter membrane bubble point values with those given by mixed esters of cellulose filters, that the retention virtues of the latter membrane will hold true for the former. This fancied relationship is especially invoked for new membrane types in an attempt to "grandfather" in the new on the authority of the old. Ignored are the facts that different polymers may well give different bubble point values, that the pore-size distributions may be vastly different, that the pore shapes are certainly different, and that the particle-adsorption propensities of the polymers also differ. No great harm is done by this technical absurdity, however, for bubble point values are, in any case uncertain guides for retention forecasting (Lukaszewicz, et al 1978) [2]. Experience is the reliable guide.

In the manufacture of microporous PTFE a lubricating agent which is a petroleum solvent (e.g., kerosine or naptha), maybe added to the finely divided resin at the preforming stage. The resin and lubricant are formed into a billet which is extruded through a die and calendared into a tape. The tape is then stretched in a controlled manner into a thin membrane, the rate of stretch being important, and is sintered, that is, taken above its melting point.

The lubricant is removed either before or after stretching, by total evaporation. In the Garlock Inc. experience, the presence of the lubricant shows itself as an obvious peak at approximately 224°C in differential scanning colorimetry. The finished microporous PTFE membrane is intended, of course, to be free of the lubricant. Figure 1 is a scanning electromicrograph of a 0.45 um-rated PTFE filter magnified 2400 times. The structural composition features elongated strands radiating from nodular regions of joining. The pores are in the nature of elongated slits; their form is quite different from that of the pores characterizing inverse phase membranes.

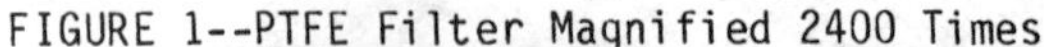

FIGURE 1--PTFE Filter Magnified 2400 Times

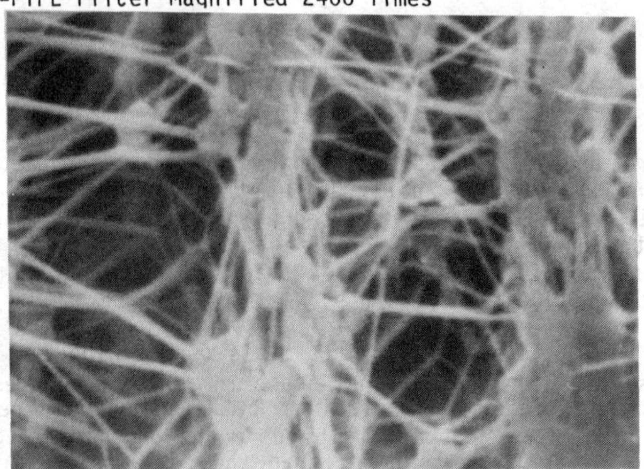

The hydrophobicity of the microporous PTFE membranes are measured by determining the pressure needed to intrude water into their pores. Physical characteristics, such as thickness and tensile properties are monitored using an electronic micrometer gauge for the former, and testing machine, such as an Instron Universal Tester, for the latter. Air flow through the microporous membrane is measured by means of a Gurley Densitometer or a Frazier Air Permeability Machine, depending upon customer requirements.

The microporous PTFE membranes are widely used for their solvent resistance in situations involving aggressive solvents, such as in crystalline antibiotics preparation but especially in semiconductor manufacture where such aggressive etching fluids as in Table 2 are used. The inertness of the PTFE filters to solvents is limited by that of the laminating substrates. In the pharmaceutical industry in accordance with the June 1, 1976 "Current Good Manufacturing Practice" in the Manufacture, Processing, Packaging, or Holding of Large Volume Parenterals, they are also widely utilized as air-vent filters, and as gas filters in general; being hydrophobic and not being composed of fibrous materials [3]. This latter property is seen as obviating the fiber-shedding problem considered inherent to fibrous air filters, those composed of cellulosic fibers, such as cotton or paper, of glass fibers, or potassium titanate fibers. The inherent hydrophobic quality of microporous PTFE is desirable. Hydrophobicity induced in other filter materials such as nylon by silicone surface coatings is often fugitive; non-performance resulting from wear or other ravages of time. Because of its inherent hydrophobic property, PTFE has a most limited attraction for water. Moisture condensation within vent filters composed of PTFE or liquid water intruded into it, will be expelled at relatively low pressures. This reduces the risk of storage-tank collapse caused by vent filter blockage due to water adsorption.

Proteins, probably through hydrogen bond formation, may adsorb to membrane surfaces. Tanny and Meltzer (1978) observed that the titer of prepared flu vaccines was diminished through the use of mixed esters of cellulose filters [4]. Other polymeric membrane materials are less absorptive; many satisfactorily so. Thus, polysulfone filters can be successfully used in plasmaphoresis work where protein concentration alteration by absorption loss must be avoided. However, proteins may absorb to PTFE and to other non-polar surfaces by hydrophobic interactions. Microporous PTFE membranes are applied in filtration applications in pharmaceuticls wherein protein loss is not of a concern such as the processing of aqueous parenterals or where absorption of charged cites is to be observed.

TABLE--2 Semiconductor Etchants and Filtrative Processing Temperatures

Etchant	Filtration Temperature
1 part Hydrofluoric Acid/ 50 parts water	Ambient to 70°C (158°F)
1 part Hydrofluoric Acid/ 10 parts water	
1 part Hydrofluoric Acid/ 4 parts water	
13 parts Ammonium Fluoride to 2 parts Hydrofluoric Acid	Ambient
Conc. Sulfuric Acid	90°-125°C (194-257°F)
5 parts water, 1 part Hydrochloric Acid 1 part Peroxide	70°-90°C (158-194°F)
4 parts Sulfuric Acid to 1 part Peroxide	150°C (302°F)

One large filter manufacturer who produces a miriad of quality less expensive polymer membranes suggests the use of PTFE filters in hot water service applications since more readily wettable membranes weaken and embrittle due to hydrolysis [5].

Although PTFE filters cannot directly be wetted by water (except under the impetus of higher pressures whereby water is intruded into the pores), they can easily be wetted by alcohols. These, if miscible with water, can be replaced by water, using aqueous flushings. Thus, a microporous PTFE filter is first wetted by an alcoholic rinse, followed by a water flush - without total alcohol evaporation from the PTFE surface having been permitted. A water layer can thereby essentially be established into a PTFE surface. Filtration of an aqueous protein preparation may then be carried out, taking advantage of the PTFE propensity not to charge-bind most proteins. Protein loss through charge-involved adsorption is thus avoided.

LIMITED USE OF FLUOROCARBON FILTERS

Although fluorocarbon filters do not possess many of the limitations of other polymer membrane types, their use in industry, to date, has been relatively limited. Albeit for certain inconveniences such as pre-wetting PTFE membranes prior to aqueous solution processing, the greatest factor in their limited use has been their higher price when compared to cellulosic, polysulfone and polyvinylidene fluoride membranes. Undoubtedly, as a result of the technological advances cited herein, prices for these filter materials will decrease making them price competitive with other polymer types. In any case the inertness and lack of extractables fluorocarbon filters offer will compensate for the premium they demand.

Telflon® is a registered trademark of E. I. Du Pont.

REFERENCES

[1] Imbalzano, J. F., Moody, J. R., and Katzer P. J., "The Effect Of Selected Aqueous Semiconductor Reagents On Commercial Piping Of Teflon® PFA Fluorocarbor Resin and Of Polyvinylidene Fluoride Fluoropolymer Resin", paper presented at Semiconductor Equipment and Materials Institute, Inc., Technical Confirence at Semicon West 5/23/85.

[2] Lukaszewicz, R. C., Tanny, G. B. and Meltzer, T. H., "Membrane-Filter Characterizations and Their Implications for Particulate Retention," Pharmaceutical Technology, Vol. 2, No. 11, 1978, pp. 76-86.

[3] "Human Drugs", Federal Register, 1976, Vol. 41 (106), 212.73.

[4] Tanny, G. B., and Meltzer, T. H., "The Dominance of Adsorptive Effects on The Filtrative Sterilization of A Flu-Vaccine", Journal Parenteral Drug Association, Vol. 32, No. 6, 1978, pp. 258-267.

[5] Catalogue 28210 MJSF, 1982, Pall Corporation, Cortland, N. Y.

Peter R. Johnston

GENERAL CONSIDERATIONS IN PERFORMING A LIQUID FILTRATION TEST

REFERENCE: Johnston, P.R., "General Considerations in Performing a Liquid Filtration Test," Fluid Filtration: Liquid, Volume II, ASTM STP 975, P.R. Johnston and H.G. Schroeder, Eds., American Society for Testing and Materials, Philadelphia, 1986.

ABSTRACT:Discussed here are three general areas of study that should be included: Permeability of the medium to the fluid of interest, Efficiency with which the medium stops different-size particles in that fluid, Life or Capacity of the medium.

KEYWORDS: permeability, filtration efficiency, filtration ratio, life or capacity of the filter medium, viscous flow.

This paper is offered to this symposium as a basis for a planned round-table discussion. The evaluation of a filter medium may begin with some measure of pore size (via a fluid-intrusion method), or porosity (ratio: void volume/bulk volume), or thickness; but, eventually that medium must be examine via an actual filtration test--using a fluid and using test particles of interest. In designing or performing a filtration test the investigator considers the following three areas of study.

PERMEABILITY OF THE MEDIUM

The investigator begins his study with the consideration of the size of the fluid stream a filter medium will be asked to clarify. Which is to say that the size (area) of the filter medium must "match" the size of the stream (volumetric flow rate) and the properties of the fluid. In a permeability measurement the investigator will make a plot of pressure drop across the two faces of the medium (the upstream face and the downstream face) versus the velocity with which the clean (particle free) liquid approaches the face. A thorough study will reach the kind of plot shown in Fig. 1.

The match between stream size and area of filter medium will be in that low velocity range in Fig. 1 where ΔP is directly proportional to v (the slope is 1.0 on this log-log plot) and where ΔP is only the result of viscous drag.

Johnston is Senior Project Engineer, Ametek, Inc., 502 Indiana Ave., Sheboygan, Wisconsin 53081

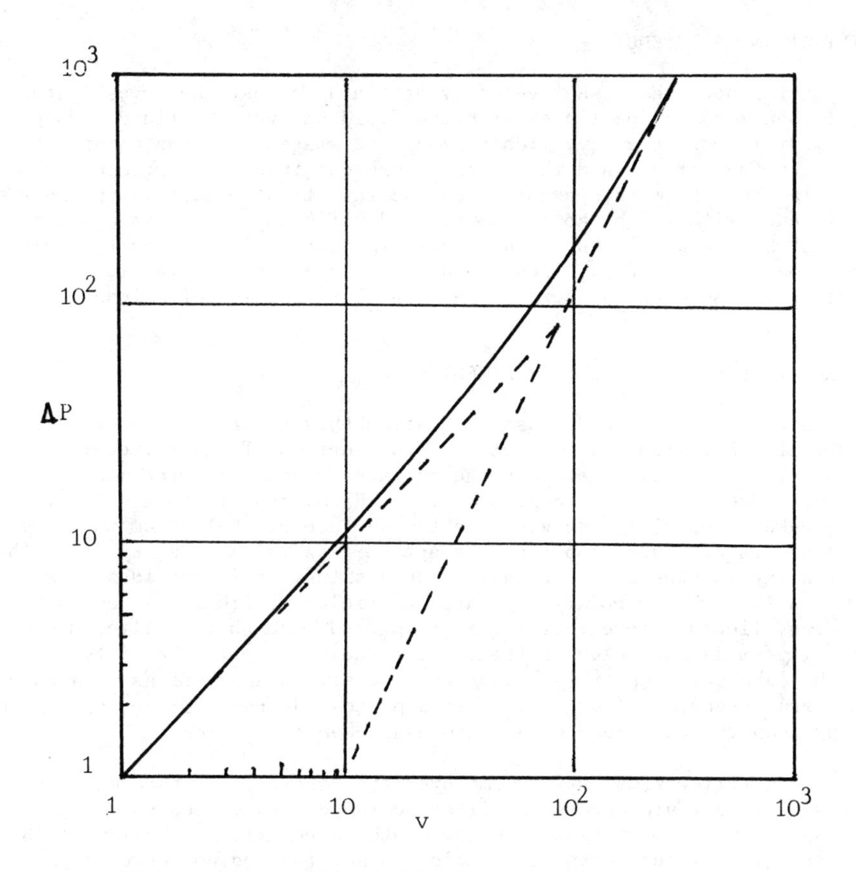

FIG. 1--Illustration of how the difference in pressure on the two separate faces of a filter medium, ΔP, changes with changes in velocity of liquid, v, approaching the medium.

That is, at low values of v where filtration is usually done, ΔP is directly proportional to v because ΔP is primarly the results of viscous drag. At high values of v, ΔP is proportional to v^2, and ΔP is primarly the results of inertial effects.

The general equation in liquid flow (discussed in ASTM F902, and ref. 1) is

$$\Delta P/z = \alpha n v + \beta \rho v^2$$

where: z = thickness of the medium
α = viscous term coefficient
n = viscosity of the liquid
β = inertial (kinetic energy) term
ρ = density of the liquid.

The corresponding way to view the flow of gas through a medium is explained in ref. 2.

FILTRATION EFFICIENCY

Having decided on what velocity of fluid to test the investigator will then mix test particles with the fluid and run the fluid through the medium. During early, middle, and late stages of a run samples of both the feed stream and the filtrate are examined for particle-size distribution. From this examination the investigator will learn the efficiency with which the medium stopped different size particles. A plot of his data will, as a rule, describe straight lines on the coordinates shown in Fig. 2. During a test run an acceptable filter medium will not show a decrease in overall filtration efficiency.

CAPACITY OR LIFE OF THE FILTER MEDIUM

Where many authors discuss filtration they refer to the situation where all particles are collected on the surface of the filter medium--this is, no particles get pass the surface to penetrate the medium or become lodged within the depths of the medium. And, in this situation, as a cake of particles develops on the surface of the medium the only increase in pressure drop (for a constant flow rate) is a result of the increasing thickness of the cake. Such a situation is the ideal case where a filter aid (relatively large particles or fibers) is added to a cloudy liquid before that liquid is passed through the filter medium. The very small particles in the cloudy liquid cling to the filter aid. As the cake builds up it is either continously scraped off as in a rotary-drum, vacuum filter, or, as in a plate-and-frame filter, filtration stops when the cake reaches a maximum-allowable thickness.

Where filter aids are not employed the particles of hard materials or "goo" in a fluid feeding a filter medium do become lodged within the depths of that medium (sometime the depth is relatively shallow) with the result that the pores become blocked and the medium looses permeability from being plugged up.

There exists four classical mathematical expressions to describe Cake filtration, Intermediate filtration (almost but not complete plugging), Standard blocking, and Complete blocking. A comprehensive collection-review of these expression--for both constant-rate and constant pressure filtration--is provided by Grace (4).

Some writers have compared their experimental plugging curves to these mathematical curves in an effort to determine the nature of the plugging process in their own specific operations. Those writers teach that in making this comparison the investigator should plot his experimental data in the form of four different graphs to see which graph shows the straightest line to indicate the specific plugging function. Alternatively, Johnston & Beals (3) offer that from normalized transformations of these four mathematical expressions one may represent all four on one log-log plot. Thus, when the investigator wants to see which plugging law seemed to prevail he merely plots his experimental data on log-log paper without regard to what units of measurements he conviently made for pressure drop, flow rate, time, or volume filtered. After making his plot he merely compares his curve to the mathematical curves to see over which he may superimpose his experimental data. These normalized mathematical expressions, and the curves they describe, are shown in Figs. 3, 4 and 5.

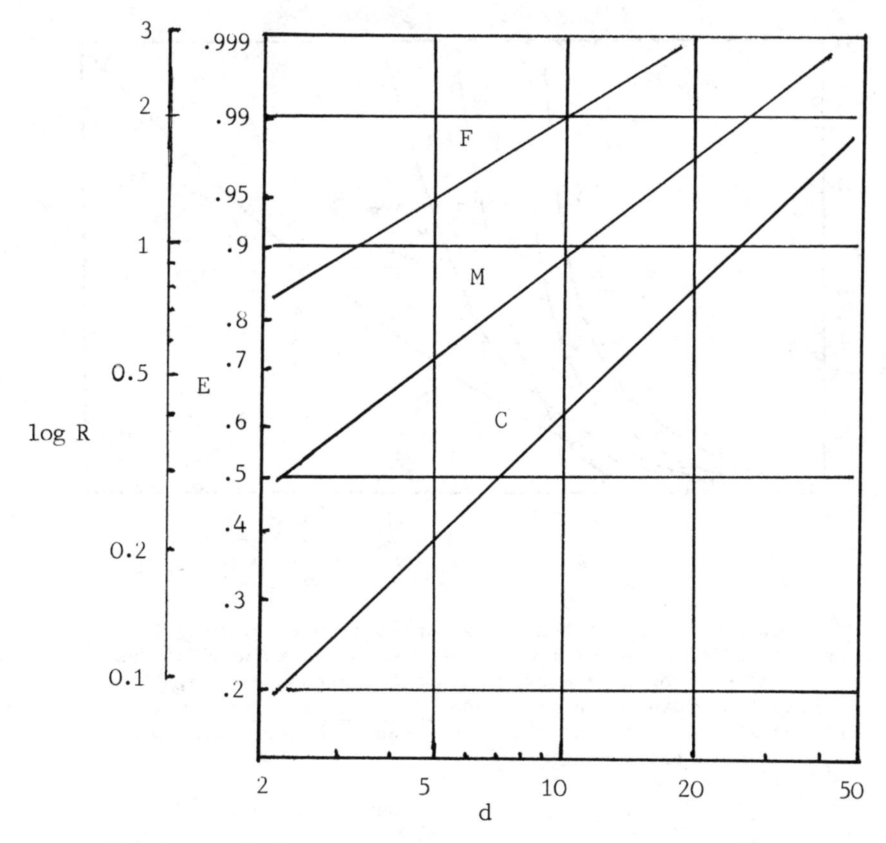

FIG. 2--Typical performance of a filter medium in liquid filtration. That is, straight lines are seen in this kind of plot--slopes may vary. d = particle diameter, µm; E = efficiency with which individual size particles are stopped by, say, a coarse-grade medium, C, a medium-grade, M, and a fine-grade, F. R = filtration ratio, 1/(1 - E).

If within the test conditions used to characterize these three filter media a less viscous liquid is used the coarse-grade medium may perform as a medium- or fine-grade one. This same change may be seen if--even with liquids of identical viscosities--an oil is used in place of water.

Liquid velocity does not appear to be an important variable, within the viscous-flow region, as it is for small particles in gas filtration.

The meaning of particle diameter, as determined via different types of automatic particle counters, is addressed in ASTM F660.

An example of how the particle-size distribution in the feed stream is seen in comparison to that in the filtrate is provided in ASTM F795 (constant-rate filtration) and in ASTM F796 (constant-pressure filtration). A single distribution can be depicted in four different ways.

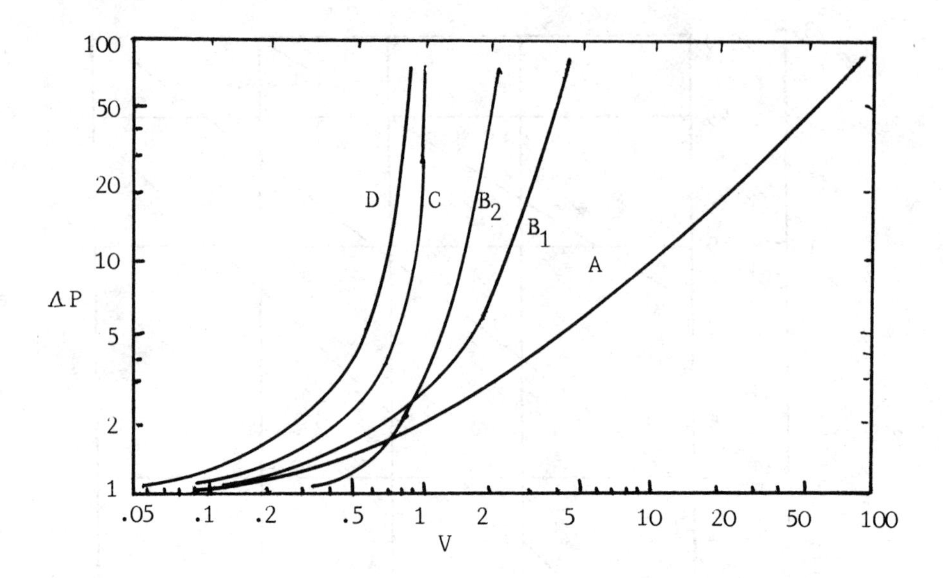

FIG. 3--Plot of mathematical expressions describing the rate at which a filter medium plugs with accumulated solids.

In these normalized expressions the pressure drop, ΔP, across the medium at a given and constant flow rate is 1.0 at the start of the run, then increases with increased volume, V, of liquid filtered.

A—$\Delta P = V + 1$, Cake filtration, max. slope = 1.0

B_1—$\Delta P = e^{V}$ ⟩ Intermediate filtration

B_2—$\Delta P = e^{V^2}$

C—$\Delta P = (1 - V)^{-2}$ Standard blocking

D -$\Delta P = (1 - V)^{-1}$ Complete blocking

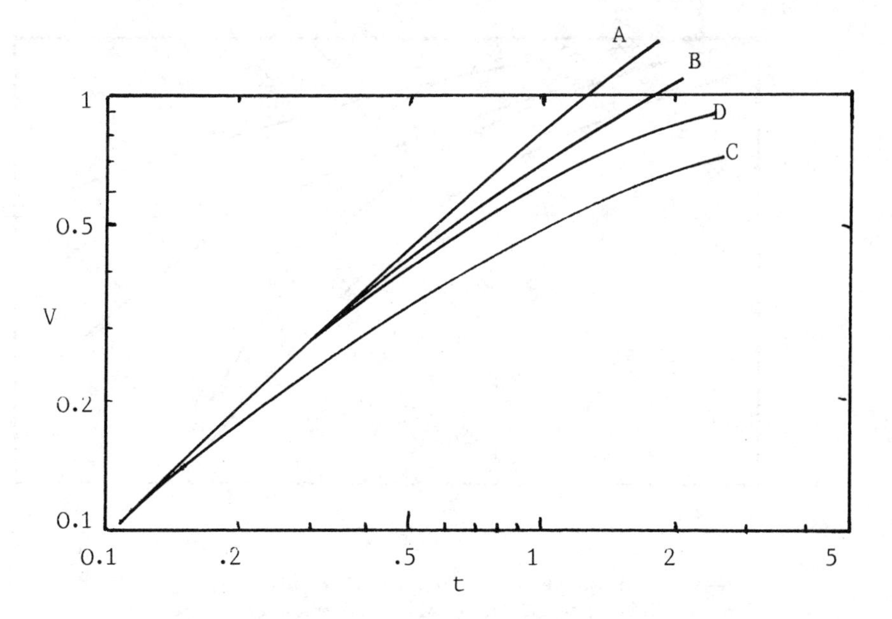

FIG. 4--Plot of mathematical expressions describing the rate at which a filter medium plugs with accumulated solids.

In these normalized expressions representing constant driving pressure and relating volume of liquid filtered, V, to time, t, the volumetric flow rate, dV/dt, initially at 1.0, falls with increased time. See alternative plots in Fig. 5.

A--$V = (4t + 4)^{1/2} - 2$, cake filtration, slope reaches 0.5

B--$V = \ln(1 + t)$, intermediate filtration

C--$V = t/(t + 1)$, standard blocking

D--$V = 1 - e^{-t}$, complete blocking

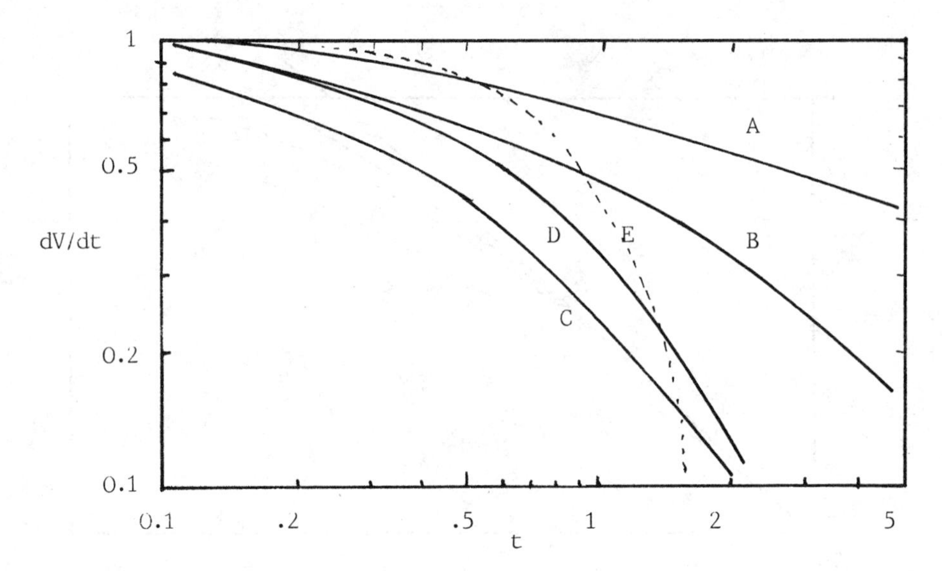

Fig. 5--Plot of mathematical expressions describing the rate at which filter media plug with accumulated solids.
In these normalized expressions representing constant driving pressure and relating volume of liquid filtered, V, to time, t, the volumetric flow rate, dV/dt, initally at 1.0 falls with increased time. See alternative plots in Fig. 4

A--$dV/dt = 1/(t + 1)^{1/2}$, cake filtration, slope reaches -0.5

B--$dV/dt = 1/(t + 1)$, intermediate filtration

C--$dV/dt = 1/(t + 1)^2$, standard blocking

D--$dV/dt = e^{-t}$, complete blocking

E--Fast plugging (three expressions in ref. 3)

REFERENCES

(1) Johnston, P.R.,"Fluid Filter Media: Measuring the Average Pore Size and the Pore-Size Distribution, and Correlation with Results of Filtration Tests," Journal of Testing and Evauation, Vol. 13, No. 4, July 1985, pp. 308-315

(2) Perry's Chemical Engineers Handbook, McGraw-Hill, New York, N.Y., 5th ed., 1973, pp. 5-54 (Porous Media)

(3) Johnston, P.R, and Beals, D.L.,"Constant-Pressure Liquid Filtration: Mathematical Models for Fast Plugging of the Filter Medium," Journal of Testing and Evaluation, Vol. 8, No. 2, March 1980, pp. 57-62

(4) Grace, H.P.,"Structure and Performance of Filter Media, Part I," American Institute of Chemical Engineers Journal, Vol. 2, No. 2, Sept 1956, pp. 307-315.

Eugene A. Ostreicher

PERFORMANCE EVALUATION OF INDUSTRIAL FILTER CARTRIDGES

REFERENCE: Ostreicher, E.A., "Performance Evaluation of Industrial Filter Cartridges", Fluid Filtration: Liquid, Volume II, ASTM STP 975, P.R. Johnston and H.G. Schroeder, Eds., American Society for Testing and Materials, Philadelphia, 1986

ABSTRACT: The selection of a cost/performance optimized filter cartridge for an industrial filter application is an empirical process. The inter-relationships between filter cartridge construction and material, fluid being filtered, contaminant type and particle size distribution, and the specific filtration results required by the user, are complex and do not allow rigorous prediction of performance for any given application. Relatively undefined single-valued characterizations of a filter cartidge's performance, such as "nominal rating", "absolute rating", and "contaminant capacity" have proven to be somewhat less than useful in this selection process.

Properly designed, conducted and interpreted laboratory performance evaluation tests can help to reduce the complexity of the selection process by providing a clear and comprehensive picture of the basic "personality" of a given filter cartridge.

KEYWORDS: fluid filters, filtration, particle removal efficiency, turbidimetric efficiency, filter performance evaluation

INTRODUCTION

As a technology matures, there is always an eventual movement to the goals of standardization and interchangeability. The achievement of such goals requires that the following conditions exist:

1. There must be a technically rigorous understanding commonly shared by both the user and the producer, of the significant performance characteristics.

Mr. Ostreicher is Manager, Filter Development at Cuno Incorporated, 400 Research Parkway, Meriden, Connecticut 06450

2. There must be standardized test methods, accepted on an industry wide basis, that allow the producer to define these performance characteristics in such a manner that intended user may intelligently select the most appropriate device, materials etc., for the intended application. Subsequently, these same test procedures may be used to verify conformance to the specified performance characteristics.

For many years, such a movement has been underway in the field of cartridge filtration. There have been numerous proposals for evaluation techniques that attempted to define the performance characteristics of such filter cartridges in technically rigorous and useful terms. Unfortunately, these have been limited, for the most part, to single point, single valued characterizations generated by test procedures that were neither commonly understood or accepted by users or producers except in very specific and limited applications. As a result, these characterizations have often resulted in more confusion (1) than enlightenment. In consequence, most successful disposable filter cartridge applications are still arrived at by lengthy, expensive, and often frustrating "trial-and-error" process.

There are a number of commonly used terms that purport to define the performance characteristics of filter cartridges. Some are based on, or imply,contaminant removal performance, as follows:

- Degree of Filtration
- Nominal Filtration Rating
- Micron Rating
- Absolute Rating
- Beta Ratio
- Alpha Ratio
- Gravimetric Efficiency
- Contaminant Capacity
- Dirt Holding Capacity

Others are based on the morphological and hydrodynamic characteristics of the filter media itself. For example:

- Permeability
- Bubble Point
- Mean Flow Pore
- Porosity
- Pore Size Distribution

Most of us in the filter industry have used these terms at one time or another in spite of the fact that there may be no commonly accepted definition of what they mean or, even more important, any great reason to believe that they even come close to allowing us to predict the potential suitability of a filter cartridge in any given application (1,2,3,4,5,6). They all fail to be useful because they are much too simplistic. They ignore the fact that filter cartridges are called upon to function in an extremely diversified range of application, fluids, comtaminants, process conditions, and user requirements and idiosyncracies. The first qroup are single point characteristics, i.e. to the extent that they are defined at all, they represent one point during the life of the cartridge. This completely ignores the fact that the filtration performance of the filter cartridge normally exhibits significant change over its life. They are singled valued, i.e., they purport a filtration efficiency for only one specific contaminant particle size distribution. As such, they ignore the fact that the user is often concerned with the removal characteristics across a broad range of particle sizes. They also ignore the fact that the filtration performance characteristics of production filter cartridges are variable within specified limits hopefully established and controlled by the producer. Lastly, and more important, the publication and attempted use of such simplistic characterizations is based on the implicit assumption that different types of filter cartridges, or similar types produced by different manufacturers, are broadly interchangable. Given the broad variety of geometries, structural configurations, raw material types and forms, and radically different processes used to produce filter cartridges, this assumption is certainly open to question. The author has spent the past thirteen years in the development of filter cartridges, and has come to view them in terms of their "personality" i.e., the specific and unique combination of overall performance characteristics that each type of cartridge exhibits.

In this paper, we will describe a basic approach to filter cartridge evaluation testing that is currently being used by Cuno. It is an attempt to broaden the performance characterization so that we can better understand the "personality" of our, or our competitors, filter cartridges. We use it to evaluate possible modifications in our raw materials, structural or geometrical characteristics, and process. We are starting to use this approach in our application engineering and find that it can be very powerful tool in minimizing the trial and error process involved in providing the user with a cost and performance optimized disposable filter cartridge. Understand that it does not eliminate the process, for each filtration application is unique and the final performance evaluation must take place in the user's actual system. It does, however, provide a much better "first" approximation and a workable technical framework to interpret the results of initial application testing and select the most appropriate candidate for subsequent testing.

FILTER CARTRIDGE PERFORMANCE CHARACTERIZATION

Figure 1 illustrates the results of a single cartridge test utilizing our current performance characterization test protocol. The cartridge tested was a cylindrical grooved resin-bonded fibre "depth type" filter cartridge (Cuno Betafine) with a "nominal filtration rating" of "one micron". The test fluid was water, at a constant flow rate of 11.4 LPM, and the contaminant was A.C. Fine Test Dust (ACFTD) at a constant concentration of 0.035 grams per liter. The test report shows, over the life of the cartridge, three performance characteristics, namely:

1. Pressure Drop vs Contaminant Added
2. Turbidimetric Efficiency vs Contaminant Added
3. Particle Removal Efficiency vs Particle Size vs Percent of Cartridge Life

This test report provides a comprehensive and complete picture of the performance of this cartridge under one very specific set of operating conditions. Additional similar tests, conducted with different fluids, flow rates, contaminant size distributions, and contaminant concentrations would give us an expanded and much more useful overall chacterization. We are on our way to understanding the "personality" of this cartridge. The first thing that we come to

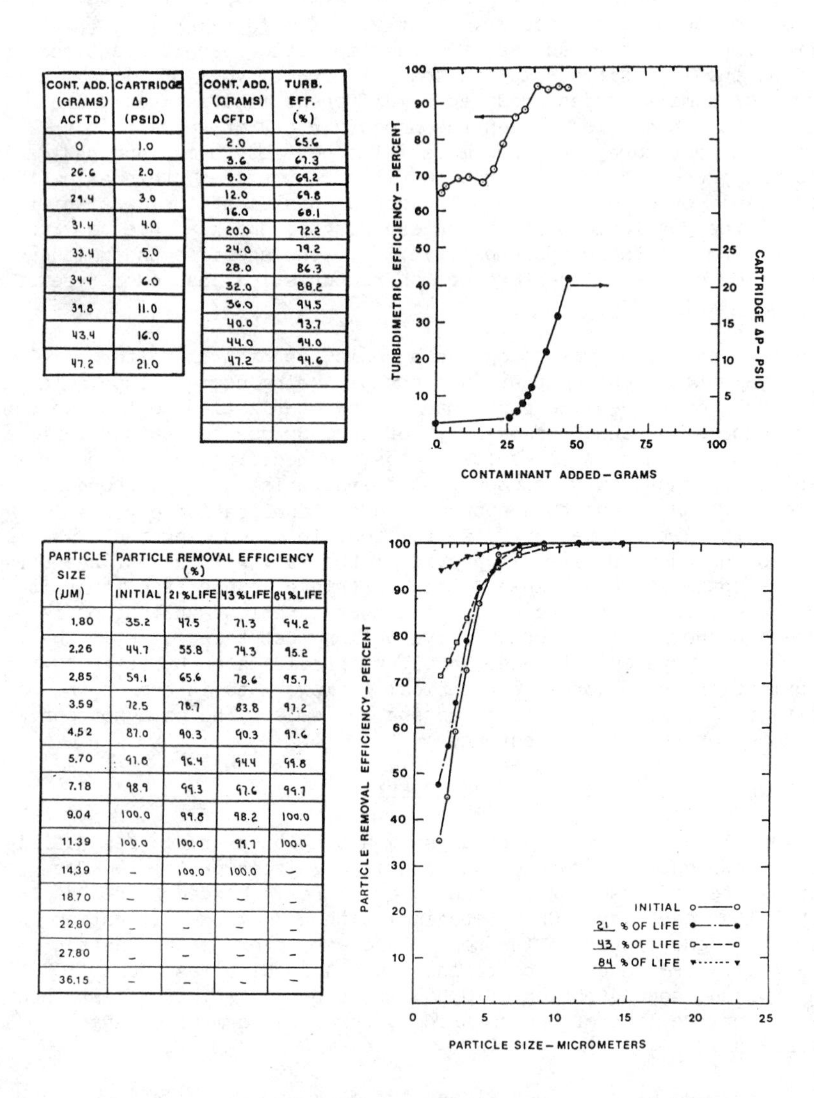

CONT. ADD. (GRAMS) ACFTD	CARTRIDGE ΔP (PSID)
0	1.0
26.6	2.0
29.4	3.0
31.4	4.0
33.4	5.0
34.4	6.0
39.8	11.0
43.4	16.0
47.2	21.0

CONT. ADD. (GRAMS) ACFTD	TURB. EFF. (%)
2.0	65.6
3.6	67.3
8.0	69.2
12.0	69.8
16.0	68.1
20.0	72.2
24.0	79.2
28.0	86.3
32.0	88.2
36.0	94.5
40.0	93.7
44.0	94.0
47.2	94.6

PARTICLE SIZE (μM)	PARTICLE REMOVAL EFFICIENCY (%)			
	INITIAL	21%LIFE	43%LIFE	84%LIFE
1.80	35.2	47.5	71.3	94.2
2.26	44.7	55.8	74.3	95.2
2.85	59.1	65.6	78.6	95.7
3.59	72.5	78.7	83.8	97.2
4.52	87.0	90.3	90.3	97.6
5.70	91.8	96.4	94.4	99.8
7.18	98.9	99.3	97.6	99.7
9.04	100.0	99.8	98.2	100.0
11.39	100.0	100.0	99.7	100.0
14.39	–	100.0	100.0	–
18.70	–	–	–	–
22.80	–	–	–	–
27.80	–	–	–	–
36.15	–	–	–	–

FIGURE 1 – FILTRATION CHARACTERISTICS – BETAFINE FILTER CARTRIDGE

understand is that the "nominal filtration rating" of "one micron" has very little value in predicting the actual filtration performance of this cartridge. As it turns out, however, this cartridge has a rather attractive "personality" compared to the "one micron nominal filtration rating" cartridge shown in Figure 2. This cartridge is a cylindrical, ungrooved melt-blown polypropylene fibre cartridge that was tested under the identical conditions. To carry the anthromorphological analogy of "personality" one step further, this second cartridge might be said to have a "schizophrenic personality disorder".

We hope that, at this point in the presentation, your interest has been sufficiently piqued to carry on with us a bit further - because we are now going to have to get a little more technical. In the introduction, it was stated that existing filter characterizations suffer from lack of definition or shared understanding of their significance. Not wishing to be guilty of repeating the same sin, it seems appropriate that we now define and discuss the functional significance of two new performance characterizations that we have presently; namely, TURBIDIMETRIC EFFICIENCY and PARTICLE REMOVAL EFFICIENCY.

TURBIDIMETRIC EFFICIENCY

The use of 90 degree light scattering turbidimetric measurements provides a simple and convenient measuring technique capable of providing a significant amount of information regarding the "real time" performance of a filter cartridge. It also is useful in terms of controlling and/or verifying the test conditions. For a given contaminant, the measured 90 degree light scattering intensity (hereafter called turbidity) is a linear function of the contaminant concentration as shown in Figure 3.

Measuring the inlet turbidity, on an intermittent or continuous basis, allows us to verify that our inlet contaminant concentration is being maintained at a constant value during the course of the test. Concurrently measuring the outlet turbidity, and comparing it with the inlet turbidity, gives us a means of characterizing an aspect of the performance of the filter. For such purpose, let us define

$$\text{TURBIDIMETRIC EFFICIENCY} = 100 \left[\frac{\text{Inlet Turbidity-Outlet-Turbidity}}{\text{Inlet Turbidity}}\right]$$

For the filter cartridge user interested in the clarification of a liquid, this characteristic can stand by itself as a significant performance parameter.

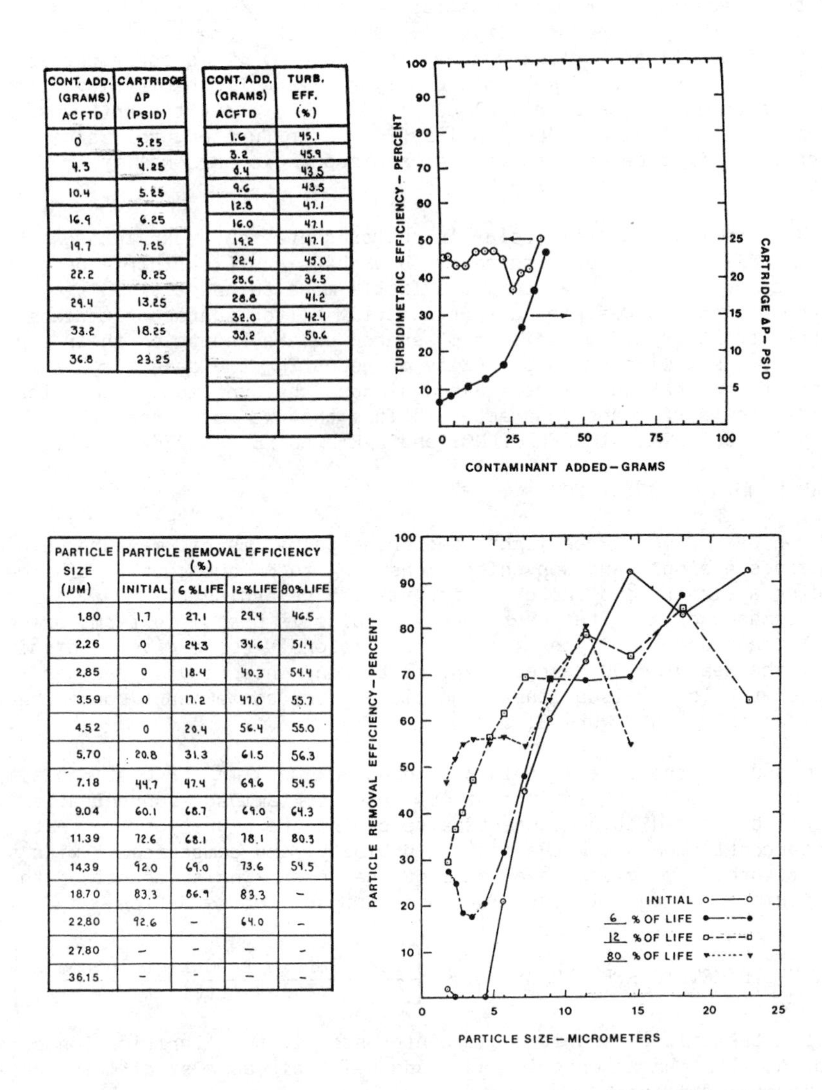

CONT. ADD. (GRAMS) AC FTD	CARTRIDGE ΔP (PSID)
0	3.25
4.3	4.25
10.4	5.25
16.9	6.25
19.7	7.25
22.2	8.25
29.4	13.25
33.2	18.25
36.8	23.25

CONT. ADD. (GRAMS) ACFTD	TURB. EFF. (%)
1.6	45.1
3.2	45.9
6.4	43.5
9.6	43.5
12.8	47.1
16.0	47.1
19.2	47.1
22.4	45.0
25.6	36.5
28.8	41.2
32.0	42.4
35.2	50.6

PARTICLE SIZE (µM)	PARTICLE REMOVAL EFFICIENCY (%)			
	INITIAL	6 %LIFE	12 %LIFE	80%LIFE
1.80	1.7	27.1	29.4	46.5
2.26	0	24.3	34.6	51.4
2.85	0	18.4	40.3	54.4
3.59	0	17.2	47.0	55.7
4.52	0	20.4	56.4	55.0
5.70	20.8	31.3	61.5	56.3
7.18	44.7	47.4	69.6	54.5
9.04	60.1	68.7	69.0	64.3
11.39	72.6	68.1	78.1	80.3
14.39	92.0	69.0	73.6	54.5
18.70	83.3	86.9	83.3	–
22.80	92.6	–	64.0	–
27.80	–	–	–	–
36.15	–	–	–	–

FIGURE 2 – FILTRATION CHARACTERISTICS – MELT BLOWN POLYPROPYLENE FILTER CARTRIDGE

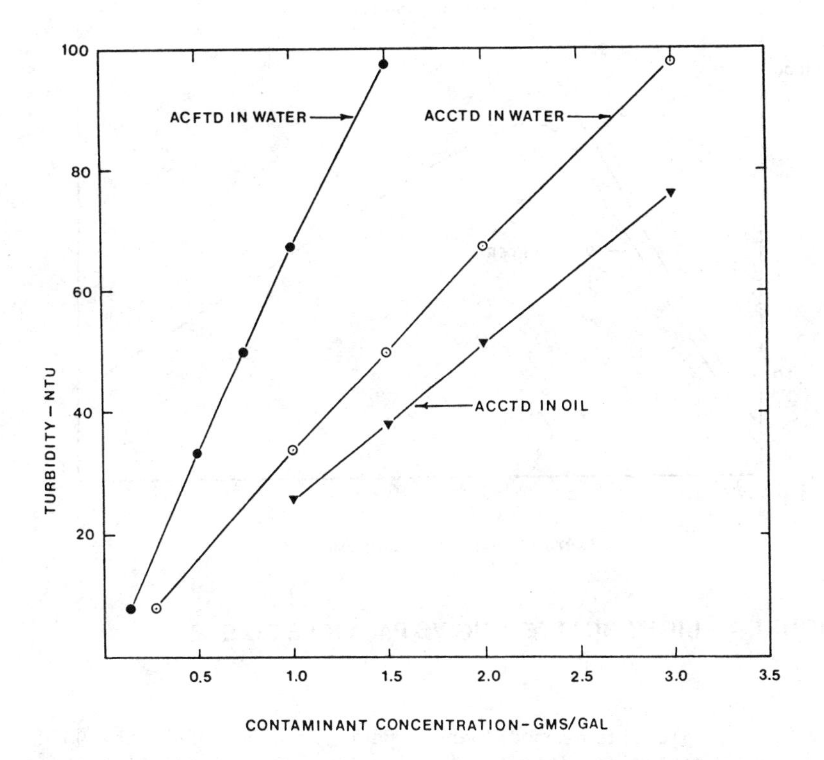

FIGURE 3—TURBIDITY VS CONTAMINANT CONCENTRATION

Unfortunately, the 90 degree light scattering characteristics are a complex function of particle size (7) as shown in Figure 4.

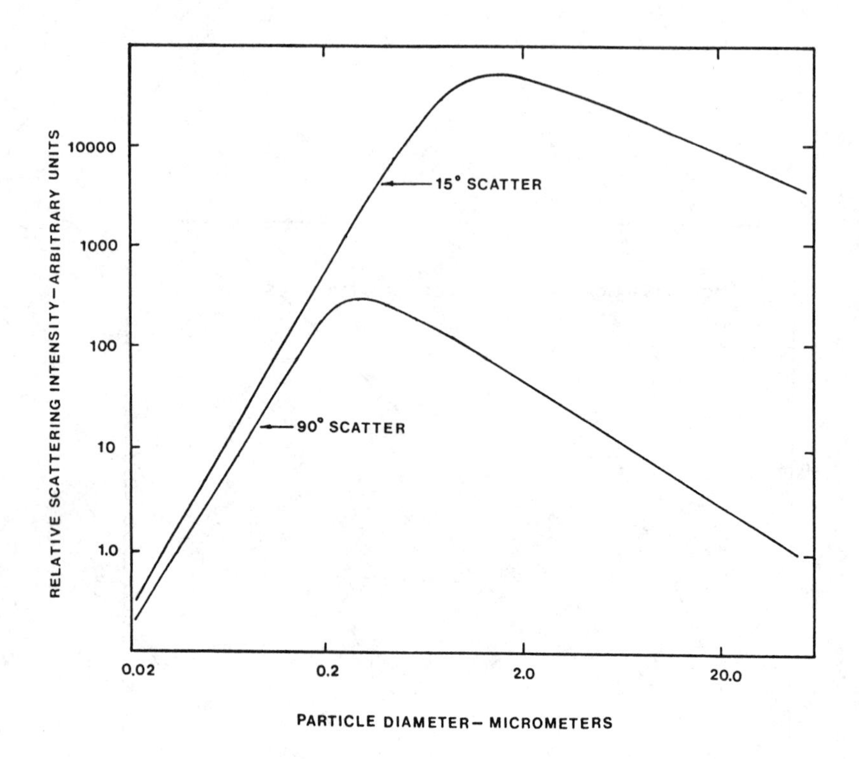

FIGURE 4—LIGHT SCATTERING VS PARTICLE SIZE

With a single size (monodisperse) particle dispersion, the 90 degree light scattering intensity is directly proportional to the number of particles per unit volume, and this has given us a powerful technique for characterizing the performance of submicronic filter media (8). Such monodisperse particles are not normally available in sizes much above 1 micrometer, however, and are much too expensive for full scale cartridge testing. As a result, we must use standard polydisperse test contaminants, such as A.C. Fine Test Dust (ACFTD) or A.C. Coarse Test Dust (ACCTD) for the testing of most industrial disposable filter cartridges. Since passage of the test contaminant dispersion thru the test cartridge changes both the concentration and particle size distribution of the contaminant dispersion, our measured TURBIDIMETRIC EFFICIENCY does not provide us with any specific data regarding either efficiency as a function of particle size (Particle Removal Efficiency) or reduction in weight concentration (Gravimetric Efficiency). As earlier investigators (9) have shown, however, it does provide us with additional information regarding the reduction in total number of contaminant particles per unit volume.

As shown in Figure 5, there is a direct log-log linear relationship between turbidity and the total number of particles per unit volume larger that one micrometer as determined by automatic particle counting techniques. Given this relationship, it would appear that our TURBIDIMETRIC EFFICIENCY gives us a definition of some sort of an "overall particle removal efficiency". Most importantly, however, it provides us with a real-time measurement that allows us to monitor the performance of the cartridge during the test and to identify anomalous behavior when it occurs. The relatively smooth, increasing withlife, turbidimetric efficiency characteristics shown in Fiqure 1 anticipates the particle removal efficiency characteristics subsequently measured for the cartridge. In contrast, the erratic turbidimetric efficiency characteristic shown in Figure 2 anticipates the erratic particle removal efficiency characteristics subquently determined for the cartridge.

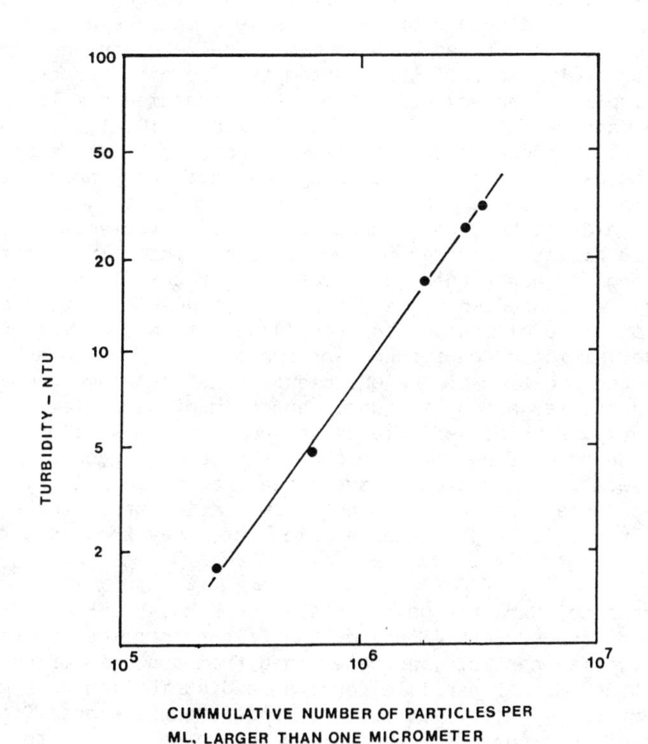

FIGURE 5 – TURBIDITY VS CONTAMINANT POPULATION

There are available commercially a broad range of instruments capable of 90 degree light scattering measurements. The simplest and least expensive type is exemplified by the Hach Model 2100 Turbidmeter (Hach Chemical Co., Ames, Iowa, U.S.A.). This instrument performs only 90 degree light scattering intensity measurements on discrete 25 ml samples of fluid. Five sensitivity ranges 0.0.2 NTU, 0-1.0 NTU, 0-10 NTU, 0-100 NTU, and 0-1000 NTU, are provided. Calibration standards with nominal values of 0.62, 10, 100 and 1000 NTU are furnished with the unit. This calibration technique is a relative one, and there appears to be no coherent system of standardization which allows calibration against an absolute standard. Fortunately, our determination of TURBIDIMETRIC EFFICIENCY depends upon the measurement of relative 90 degree light scattering intensity and the linearity of the instrument is sufficient to insure adequate accuracy for such relative measurements.

PARTICLE REMOVAL EFFICIENCY

The advent of automatic particle counters opened up a whole new realm of possibilities in terms of filter cartridge characterization. Within the industry, a number of proposals were made regarding the use of particle counter generated data to characterize cartridge performance. Unfortunately, most of these attempts were aimed at creating relatively simplistic rating definitions that presumed simple, predictable and mathematically modelable performance (10,11,12). Our experience with such characterizations was that, by the time one had "massaged" the data to fit the preconceived model, they provided very little real information regarding the performance of the cartridge. As such, they did not help us to understand the performance of our own disposable filter cartridge product line sufficiently to either properly control or apply these products in critical filtration applications. In consequence, we determined to develop a test and and evaluation protocol that would provide us with a much more comprehensive definition of cartridge performance. Of particular interest was the question of how efficiently the cartridge removed contaminant particles across the whole spectrum of particle sizes and, more importantly, how this efficiency varied over the life of the cartridge. The type of performance characterizations resulting from our test and evaluation protocol have already been shown in Figures 1 and 2. These characterizations do not show simple, predictable and mathematically modelable performance - in fact, they show just the opposite.

Our performance characterization "PARTICLE REMOVAL EFFICIENCY" defines the instantaneous efficiency of the filter cartridge in removing a specific size particle, as determined from analysis of concurrent inlet and outlet particle count/size distribution data generated by a particle counter. In order to understand the specific significance of this characterization, we are going to have to take a little trip into the intricate world of particle counter data reduction. There are two basic types of automatic particle counters; namely, the electrical resistance or electrozone type (Coulter) and the optical or light extinction type (Hiac-Royco). For those interested in the theoretical and operational considerations involved in selecting and using these counters, there is a significant literature available (13,14,15,16,17,18). Regardless of the sensing method, both types are similar in that they provide us with specific counts (particles per milliliter) of the particles within each of a number of predetermined size ranges. Table 1 exhibits "typical" data for a hypothetical single set of concurrent inlet and outlet samples as determined using a Coulter Counter. This counter has 16 output channels with each being preset to count the total number of particles in the sample having a "diameter" between a lower size limit and an upper size limit. In Table 1 for example, channel 1 is the channel that has been preset to count all of the voltage pulses of a magnitude that represents particle equal to or larger than 1.26 micrometers but smaller than 1.59 micrometers. As shown in the columns labeled "Diff." for the inlet and outlet samples, within the size range $1.26\ \mu m \leq d < 1.59\ \mu m$, there were 4299 particles/ml in the inlet sample and 2376 particles/ml in the outlet sample. Since we have inlet and outlet particle counts for the specified size range, we can certainly consider defining a "range efficiency" such that we are enabled to state the following:

The efficiency of removal for the particle size range 1.26µm to 1.59µm is equal to:

$$\text{Eff.} = 100\left[\frac{4299 - 2376}{4299}\right] = 44.7\%$$

If we carry our thinking a bit further, however, we can modify this approach to arrive at a much more useful and satisfying characterization, as follows:

1. The calculated "Size range" efficiency of 44.7% is too high a value for 1.26µm particles.

2. It is too low for 1.59µm particles.

3. Therefore, 44.7 percent must be the precise and specific particle removal efficiency for some intermediated diameter (D) particle, namely the median particle diameter within the the population of particles larger than 1.26µm and smaller than 1.59µm.

Precise determination of this median diameter requires knowledge of the particle size distribution within the size range in question. We are quite satisfied to approximate this diameter by using the arithmetic mean for the size range in question, i.e. 1.43 micrometers. We can similarly calculate efficiencies for each of the other size ranges as shown in Table 1. The resulting "PARTICLE REMOVAL EFFICIENCY" characteristic is shown in Figure 6.

If we take several sets of inlet and outlet samples at appropriate points during the life of the test cartridge, we can generate a family of particle removal efficiency curves that put us well on our way towards defining the "personality" of the cartridge under test. This approach was first proposed in 1975 by Lloyd and Ward (19) and, given the practical and theoretical ambiquities surrounding the precise significance of various methods of particle size characterzation, is at least sufficiently precise for our purpose.

CHANNEL	DIAMETER (μm)	INLET COUNTS DIFF.	INLET COUNTS CUMM.	OUTLET COUNTS DIFF.	OUTLET COUNTS CUMM.	BETA RATIO. β_x	$\overline{D}$ (μm)	PART. REM. EFFICIENCY (%)
1	1.26	4298.9	58596.5	2375.7	11504.0	5.09	1.43	44.7
2	1.59	6385.3	54297.6	2884.7	9128.3	5.95	1.80	54.8
3	2.00	7740.5	47912.3	2586.7	6243.6	7.67	2.26	66.6
4	2.52	8086.9	40171.8	1808.7	3659.9	10.98	2.85	77.6
5	3.17	7798.2	32084.9	1054.6	1848.2	17.36	3.59	86.5
6	4.00	6778.0	24286.7	499.7	793.6	30.60	4.52	92.6
7	5.05	5505.0	17508.7	200.5	293.9	59.57	5.70	96.4
8	6.35	4177.1	12003.7	67.9	93.4	128.5	7.18	98.4
9	8.00	2975.2	7826.6	19.5	25.5	306.9	9.04	99.3
10	10.08	1991.5	4851.4	4.8	6.0	808.6	11.39	99.8
11	12.70	1256.4	2959.9	1.0	1.2	2383	14.35	99.9
12	16.00	753.7	1603.5	0.17	0.2	8018	18.10	99.98
13	20.2	416.6	849.8	0.03	0.03	28327	22.8	99.99
14	25.4	224.4	433.2	0	0	-	-	-
15	32.0	132.5	208.8	0	0	-	-	-
16	43.0	76.3	76.3	0	0	-	-	-

TABLE 1 — PARTICLE COUNTER DATA REDUCTION

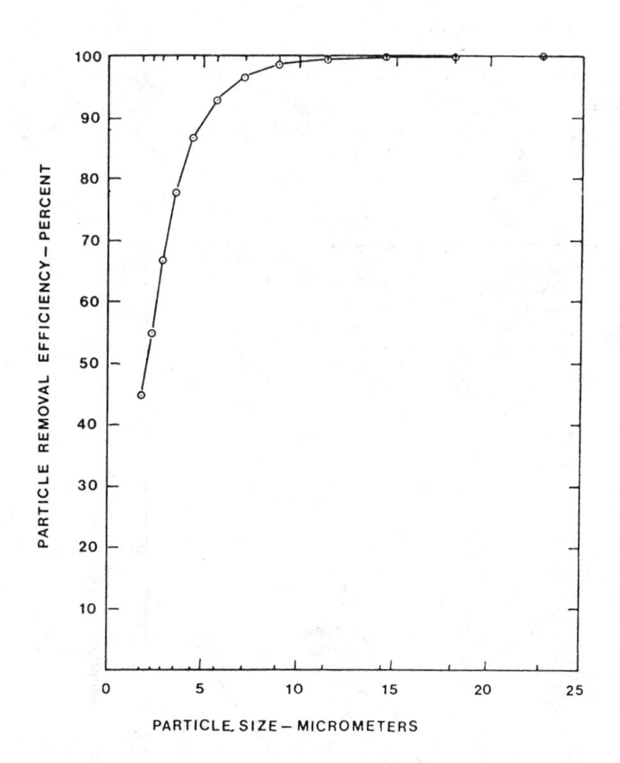

FIGURE 6 – PARTICLE REMOVAL EFFICIENCY CHARACTERISTICS FROM TABLE 1

DISCUSSION OF TYPICAL RESULTS

The first thing that application of these test methodologies lets us define is the nature of the test itself. In establishing required test conditions, we have to define and select the significant test parameters. The primary ones are:

1. Fluid
2. Flow Rate
3. Contaminant Type
4. Contaminant Concentration

If we change any of these parameters, the performance of the cartridge will also change. For example, let us look at the choice of fluid.

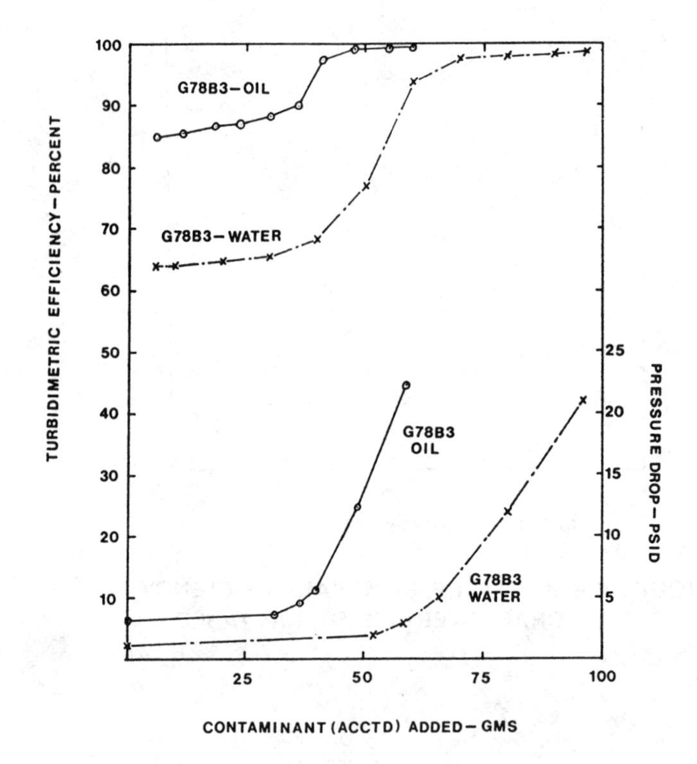

FIGURE 7—EFFECT OF FLUID—MK G78B3

Figure 7 shows the turbidimetric efficiency and pressure drop characteristics of two essentially identical cylindrical grooved resin-bonded fibre "depth-type" filter cartridges (Cuno Microklean G78B3), tested with ACCTD at a flow rate of 3.0 GPM, using water and oil as the test fluids. As we can see, this type of cartridge exhibits a higher turbidimetric efficiency and a shorter life in oil than it does in water. If we recognize the fact that the phenolic resin binder used in this cartridge produces a slightly hydrophobic surface that wets more easily with oil than with water, we probably have a reasonable explanation for the difference. Similarly, hydrophilic materials such as unresinated cellulose (cotton roving in wound cartridges, wood pulp in unresinated paper) exhibit higher efficiences and lower life in water than in oil. Part of our conception of a filter cartridge's "personality" must deal with that particular cartridge's surface wetting, absorbancy, and possible swelling characteristics in the fluid being filtered.

As another example, let us look at the choice of test contaminants. Figure 8 shows the turbidimetric efficiency and pressure drop characteristics of two essentially identical cylindrical grooved resin-bonded fibre "depth-type" filter cartridges (Cuno MicroKlean G78Y8) tested in water at 3.0 GPM using AC Coarse and AC Fine test dusts. In this case, the turbidimetric efficiency is significantly higher for the ACCTD, but, surprisingly, there is very little difference in the cartridge life between the two contaminants.

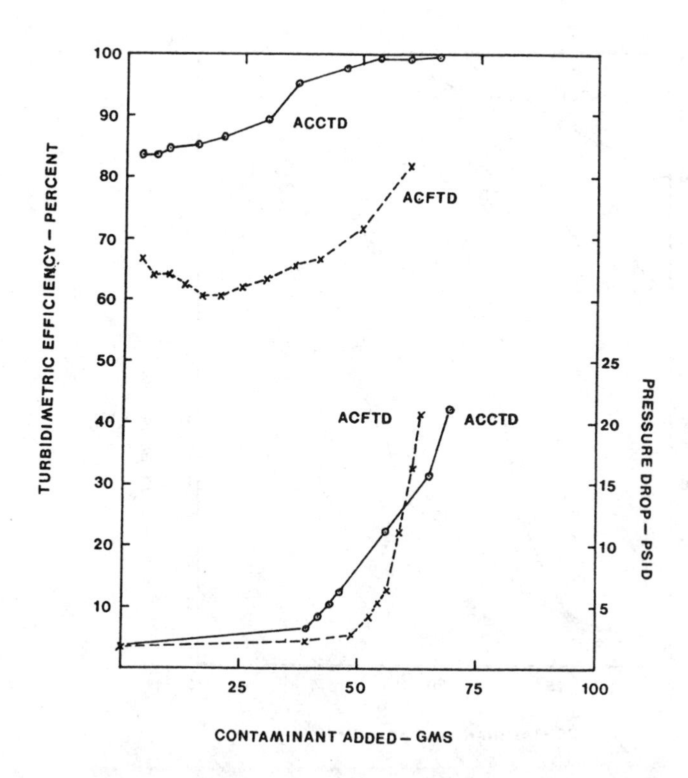

FIGURE 8 – EFFECT OF CONTAMINANT – MK G78Y8

In Figure 9, we have repeated the same test conditions with two essentially identical inserted media wound cotton cartridges (Cuno MicroWynd D-CPPY). As with the Micro Klean, the turbidimetric efficiency is higher for the ACCTD than it is for the ACFTD. Very surprisingly, the cartridge life is shorter with the finer contaminant. On the basis of a superficial analysis of the situation, one would normally expect a finer contaminant to result in a combination of reduced efficiency and increased life compared to a coarser contaminant. Almost without exception, however, we found just the opposite when utilizing these two contaminants to test cartridges in the "25µm nominal and under" range. In the case of the Micro Wynd, the finer ACFTD appears to penetrate into the graded density, resulting in a more rapid plugging. With the ACCTD, on the other hand, we stop enough of the contaminant at the outside surface of the cartridge to eventually form a cake and this has the effect of extending life. It appears, therefore, that the Micro Klean cartridge is much less sensitive to variations in contaminant particle size distribution than the Micro Wynd. Again, we have identified another facet in the "personalities" of these two cartridges. For a "10µm nominal" pleated paper type cartridge, our results show that the life for ACFTD (Figure 10) is approximately one half the life for ACCTD (Figure 11). For these cartridges, both the turbidimetric efficiency and particle removal

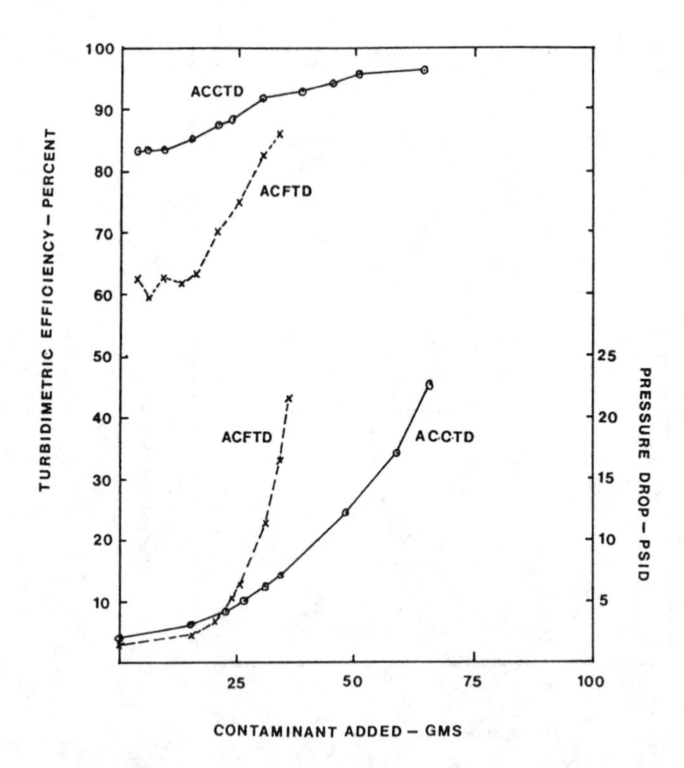

FIGURE 9 – EFFECT OF CONTAMINANT – MW D-CCPY

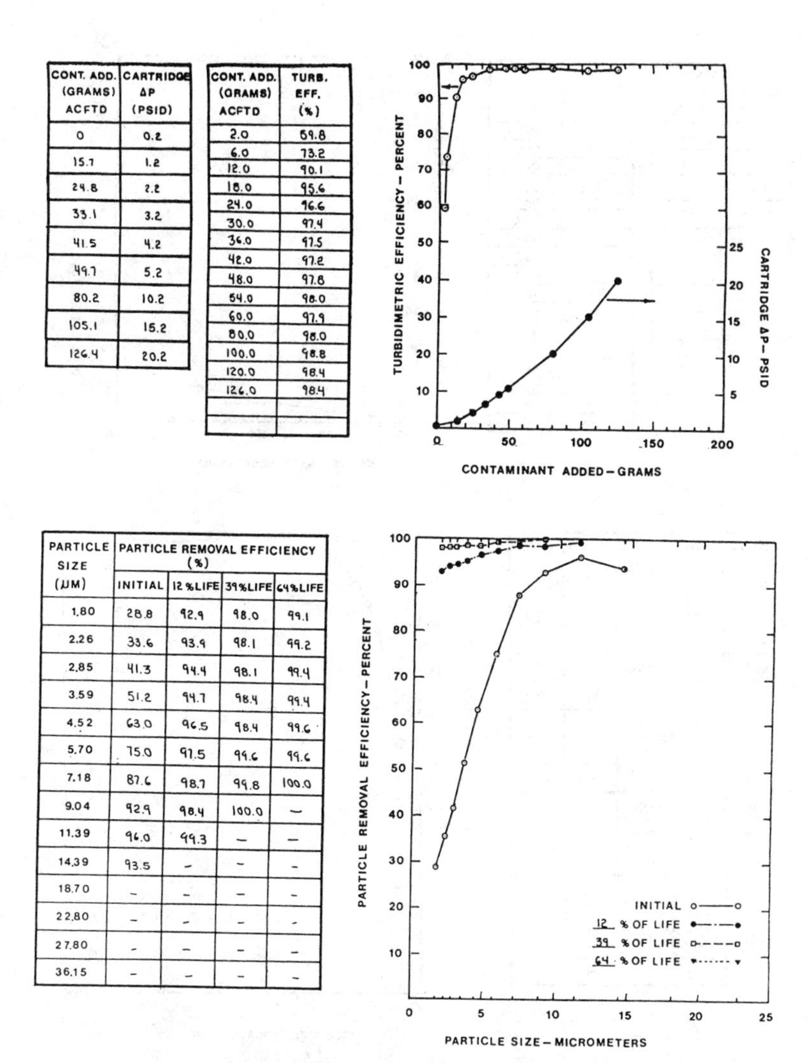

CONT. ADD. (GRAMS) ACFTD	CARTRIDGE ΔP (PSID)
0	0.2
15.7	1.2
24.8	2.2
33.1	3.2
41.5	4.2
49.7	5.2
80.2	10.2
105.1	15.2
126.4	20.2

CONT. ADD. (GRAMS) ACFTD	TURB. EFF. (%)
2.0	59.8
6.0	73.2
12.0	90.1
18.0	95.6
24.0	96.6
30.0	97.4
36.0	97.5
42.0	97.2
48.0	97.8
54.0	98.0
60.0	97.9
80.0	98.0
100.0	98.8
120.0	98.4
126.0	98.4

PARTICLE SIZE (μM)	PARTICLE REMOVAL EFFICIENCY (%)			
	INITIAL	12 %LIFE	39%LIFE	64%LIFE
1.80	28.8	92.9	98.0	99.1
2.26	33.6	93.9	98.1	99.2
2.85	41.3	94.4	98.1	99.4
3.59	51.2	94.7	98.4	99.4
4.52	63.0	96.5	98.4	99.6
5.70	75.0	97.5	99.6	99.6
7.18	87.6	98.7	99.8	100.0
9.04	92.9	98.4	100.0	–
11.39	96.0	99.3	–	–
14.39	93.5	–	–	–
18.70	–	–	–	–
22.80	–	–	–	–
27.80	–	–	–	–
36.15	–	–	–	–

FIGURE 10– FILTRATION CHARARACTERISTICS–"10 MICRON NOMINAL" PLEATED PAPER FILTER CARTRIDGE TESTED WITH AC FINE TEST DUST

CONT. ADD. (GRAMS) ACCTD	CARTRIDGE ΔP (PSID)
0	0.5
53.0	1.5
85.7	2.5
117.5	3.5
142.0	4.5
162.6	5.5
218.1	10.5
243.9	15.5
260.3	20.5

CONT. ADD. (GRAMS) ACCTD	TURB. EFF. (%)
3	72.7
6	77.6
9	82.9
15	91.7
21	94.6
33	97.2
45	95.5
60	93.0
75	97.9
90	98.8
105	98.8
120	98.6
150	99.0
180	98.5
210	98.7
240	98.1
255	98.7

PARTICLE SIZE (μM)	PARTICLE REMOVAL EFFICIENCY (%)			
	INITIAL	9 %LIFE	20%LIFE	84%LIFE
1.80	62.7	94.0	96.4	98.5
2.26	71.4	95.4	97.7	98.9
2.85	78.4	96.4	98.2	99.1
3.59	84.4	97.6	98.7	99.3
4.52	89.7	98.3	98.7	99.3
5.70	94.5	98.9	99.2	99.3
7.18	97.4	99.5	99.5	100.0
9.04	99.3	99.9	99.7	–
11.39	99.5	99.9	–	–
14.39	–	100.0	99.2	–
18.70	97.7	–	99.1	–
22.80	100.0	–	100.0	–
27.80	–	–	–	–
36.15	–	–	–	–

FIGURE 11 – FILTRATION CHARACTERISTICS – "10 MICRON NOMINAL" PLEATED PAPER FILTER CARTRIDGE TESTED WITH AC COARSE TEST DUST

efficiency characteristics show that there is a very rapid cake formation and, once the cake has formed, the remaining life and efficiency characteristics are determined by the intrinsic nature of the cake. Since the ACFTD form tighter, more easily blinded cake, it gives the noted shorter life. This specific response to contaminant particle size distribution is a typical "personality trait" for high efficiency pleated cartridges. This type of response is responsible for occasionally ambiguous results in multiple stage filtration systems, i.e., improperly chosen prefilters can fail to elicit an improvement in final filter life because they modify the original contaminant particle size distribution into one that is actually more difficult for the final filter cartridges to handle. In high solids pigmented systems, such as paints, magnetic oxide slurries, and other similar coatings, cartridges exhibiting this type of response may show erratic life characteristics in response to relatively subtle changes in pigment size distribution.

Having taken a brief look at the effect of fluid and contaminant distribution as they impact on test performance (and, by implication, on "real life" performance), let's take a quick look at what our test protocol is capable of telling us about the effect of subtle structural or material differences in filter cartridges. For example, let us look at conventional wound cartridges. These cartridges are relatively easy to produce with minimal capital investment and, as a result, are produced on a world wide basis by a large number of manufactures. It has become common practice in the industry to equate the wind pattern with a filtration rating and, indeed, many users specify their procurements on this basis. Figures 12 and 13, show the performance characterizations for two conventional polypropylene wound cartridges each having 27 x 91 wind pattern. The only difference between the two cartridges is the source of the polypropylene roving used to wind the cartridges. Superficially, both of these rovings met the same procurement specification. In spite of this, they certainly do exhibit radically different performance characteristics, with the test results showing significant differences in turbidimetric efficiency, particle removal efficiency, and life. In general, we have found that conventional wound cartridge tend to exhibit a rather broad variance in performance characteristics because of their sensitivity to roving quality and consistency. The potential user of such cartridges should recognize that the wind pattern, per se, is no indicator of specific performance, and should not assume that different manufacturers cartridges of the same wind pattern are equivalent. This is, if you will, part of the conventional wound cartridge "personality".

Each type of industrial filter cartridge represents a specific combination of process and materials that uniquely determines the basic nature of the "as-manufactured" porous structure and perhaps more importantly, the interactive response of that structure to the fluid, contaminant, and operating conditions in a specific application. Structural deformation of the porous filter media, both on a macro and micro scale, appears to be the primary cause of the idiosyncratic performance shown in Figures 2,12,13,15 and 16. In addition to dealing with the instantaneous contaminant challenge, the cartridge must be capable of retaining the "inventory" of contaminant particles that it has already removed from the fluid stream. If there is

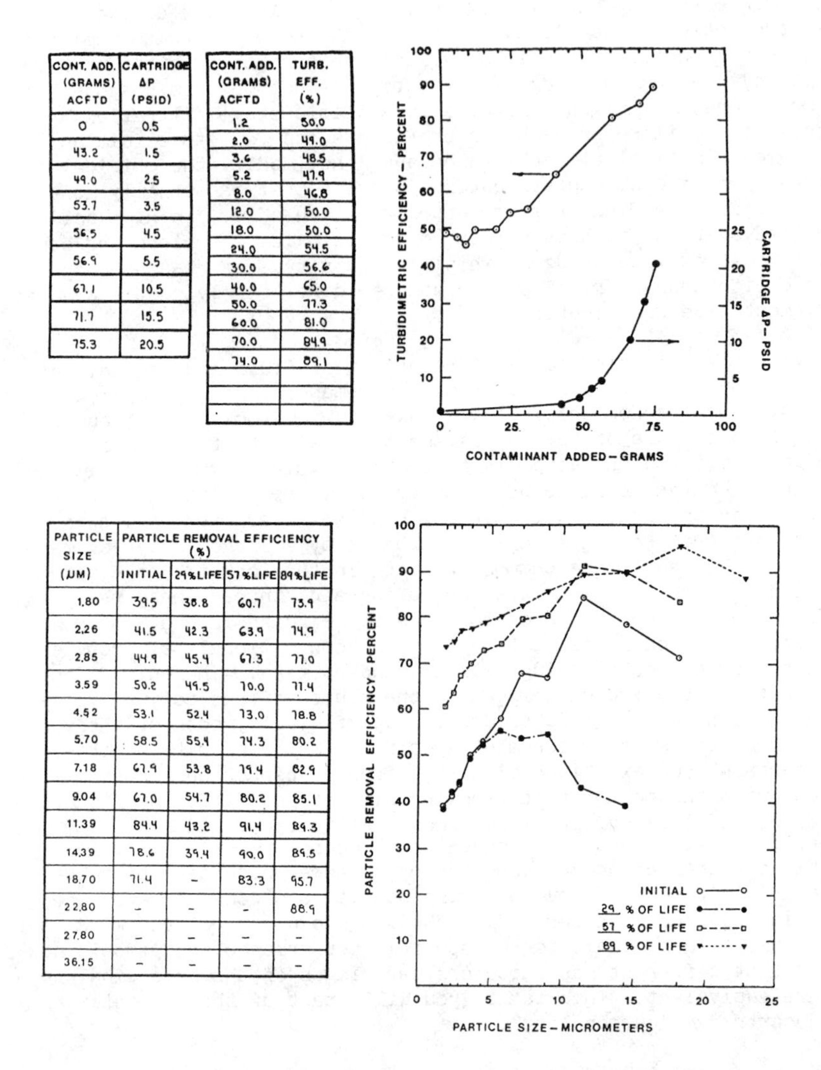

CONT. ADD. (GRAMS) ACFTD	CARTRIDGE ΔP (PSID)
0	0.5
43.2	1.5
49.0	2.5
53.7	3.5
56.5	4.5
56.9	5.5
67.1	10.5
71.7	15.5
75.3	20.5

CONT. ADD. (GRAMS) ACFTD	TURB. EFF. (%)
1.2	50.0
2.0	49.0
3.6	48.5
5.2	47.9
8.0	46.8
12.0	50.0
18.0	50.0
24.0	54.5
30.0	56.6
40.0	65.0
50.0	77.3
60.0	81.0
70.0	84.9
74.0	89.1

PARTICLE SIZE (µM)	PARTICLE REMOVAL EFFICIENCY (%)			
	INITIAL	29%LIFE	57%LIFE	89%LIFE
1.80	39.5	38.8	60.7	73.9
2.26	41.5	42.3	63.9	74.9
2.85	44.9	45.4	67.3	77.0
3.59	50.2	49.5	70.0	77.4
4.52	53.1	52.4	73.0	78.8
5.70	58.5	55.4	74.3	80.2
7.18	67.9	53.8	79.4	82.9
9.04	67.0	54.7	80.2	85.1
11.39	84.4	43.2	91.4	89.3
14.39	78.6	39.4	90.0	89.5
18.70	71.4	–	83.3	95.7
22.80	–	–	–	88.9
27.80	–	–	–	–
36.15	–	–	–	–

FIGURE 12 – FILTRATION CHARACTERISTICS–CONVENTIONAL WOUND FILTER CARTRIDGE–27 X 91 WIND SUPPLIER "A" POLYPROPYLENE ROVING

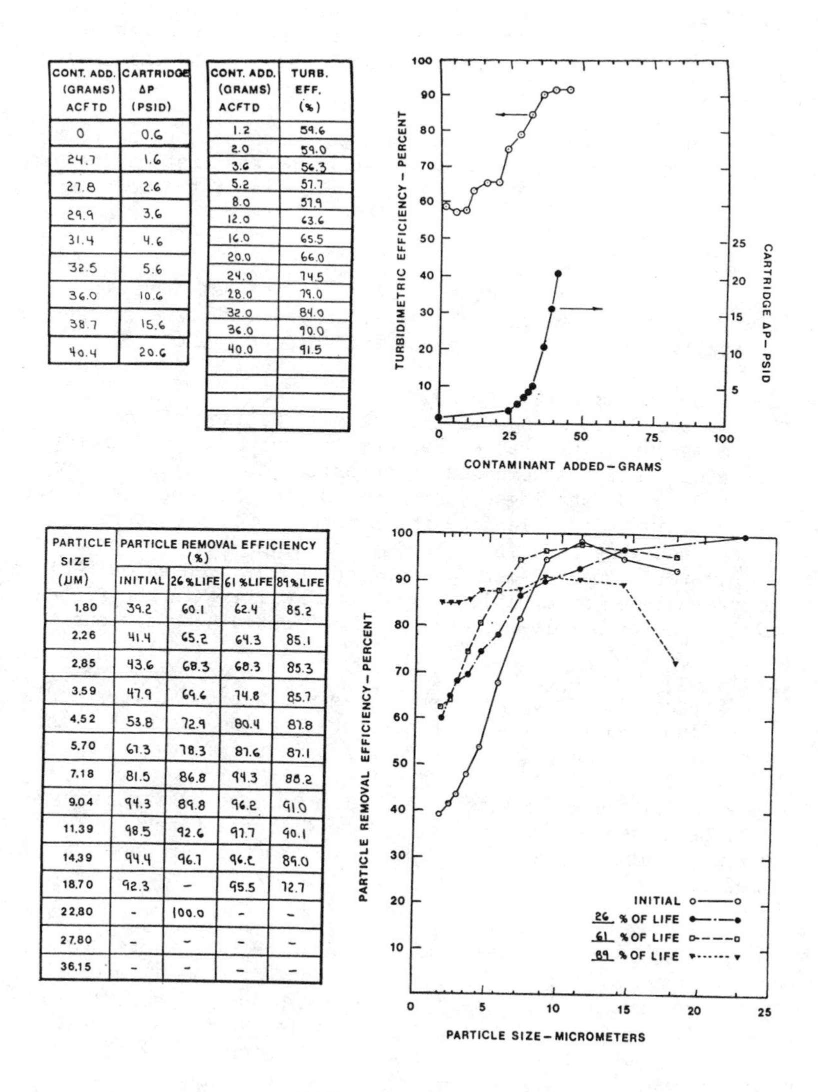

CONT. ADD. (GRAMS) ACFTD	CARTRIDGE ΔP (PSID)
0	0.6
24.7	1.6
27.8	2.6
29.9	3.6
31.4	4.6
32.5	5.6
36.0	10.6
38.7	15.6
40.4	20.6

CONT. ADD. (GRAMS) ACFTD	TURB. EFF. (%)
1.2	59.6
2.0	59.0
3.6	56.3
5.2	57.7
8.0	57.9
12.0	63.6
16.0	65.5
20.0	66.0
24.0	74.5
28.0	79.0
32.0	84.0
36.0	90.0
40.0	91.5

PARTICLE SIZE (µM)	PARTICLE REMOVAL EFFICIENCY (%)			
	INITIAL	26 % LIFE	61 % LIFE	89 % LIFE
1.80	39.2	60.1	62.4	85.2
2.26	41.4	65.2	64.3	85.1
2.85	43.6	68.3	68.3	85.3
3.59	47.9	69.6	74.8	85.7
4.52	53.8	72.9	80.4	87.8
5.70	67.3	78.3	87.6	87.1
7.18	81.5	86.8	94.3	88.2
9.04	94.3	89.8	96.2	91.0
11.39	98.5	92.6	97.7	90.1
14.39	94.4	96.7	96.2	89.0
18.70	92.3	-	95.5	72.7
22.80	-	100.0	-	-
27.80	-	-	-	-
36.15	-	-	-	-

FIGURE 13 – FILTRATION CHARACTERISTICS – CONVENTIONAL WOUND FILTER CARTRIDGE – 27X91 WIND SUPPLIER "B" POLYPROPYLENE ROVING

significant deformation of the filter media, in response to increasing differental pressure, initially retained contaminant particles may be released in the form of agglomerates. For the typical flat end sealed cartridge, axial compression may result in loss of sealing with subsequent bypass of contaminant. It is very worthwhile to perform cartridge testing in a clear, transparent plastic housing so that any gross deformation may be observed.

In addition to structural response to differential pressure, there are a number of other possible interactive responses that may have a significant impact on cartridge performance. Some of the more important ones are as follows:

1. Electrokinetic Effects
   The interactive effects of fluid ionic chemistry, pH, contaminant surface charge and filter media surface charge can radically impact upon filter cartridge performance.(20)

2. Swelling of the Filter Media
   If the fluid being filtered causes swelling of the filter media material, filtration performance can be significantly effected. Filters composed of unresinated cellulosic fibres (cotton wound cartridges, unresinated pleated paper, etc) will exhibit higher efficiencies and shorter life in filtering water than they will in filtering non-polar fluids such as oils, etc. Fine polymer fibre based media, such as melt-blown polypropylene can exhibit significant swelling in the filtration of certain solvents such as chlorinated hydrocarbons.

3. Wetting
   If the fluid being filtered does not wet the filter media, air displacement from the finer pores may be inhibited, and initial filtration performance can be effected.

The possibility of such interactive effects must be evaluated and, if present, dealt with in the design of the filtration tests. Correspondingly, they must be considered in any attempt to extrapolate from a given set of test results to the conditions present in a proposed application.

Finally, there is the basic task of characterizing cartridge performance. We have already had a chance to look at some "1 micron nominal" filter cartridges; namely, the resin bonded (Figure 1) and melt blown polypropylene (Figure 2). To complete the disclosure, let us add the characterizations for a few more "1 micron nominal" cartridges, as follows:

Figure 14 - Inserted-Media Wound Cartridge (Cuno M.W. D-CCPY)

Figure 15 - Conventional Wound Cotten (39 wind)

Figure 16 - Resin Bonded Glass Fibre

Each of these characterizations demonstrates a different kind of "personality" and each certainly identifies the fact that the "1 micron nominal" rating has little basis in reality.

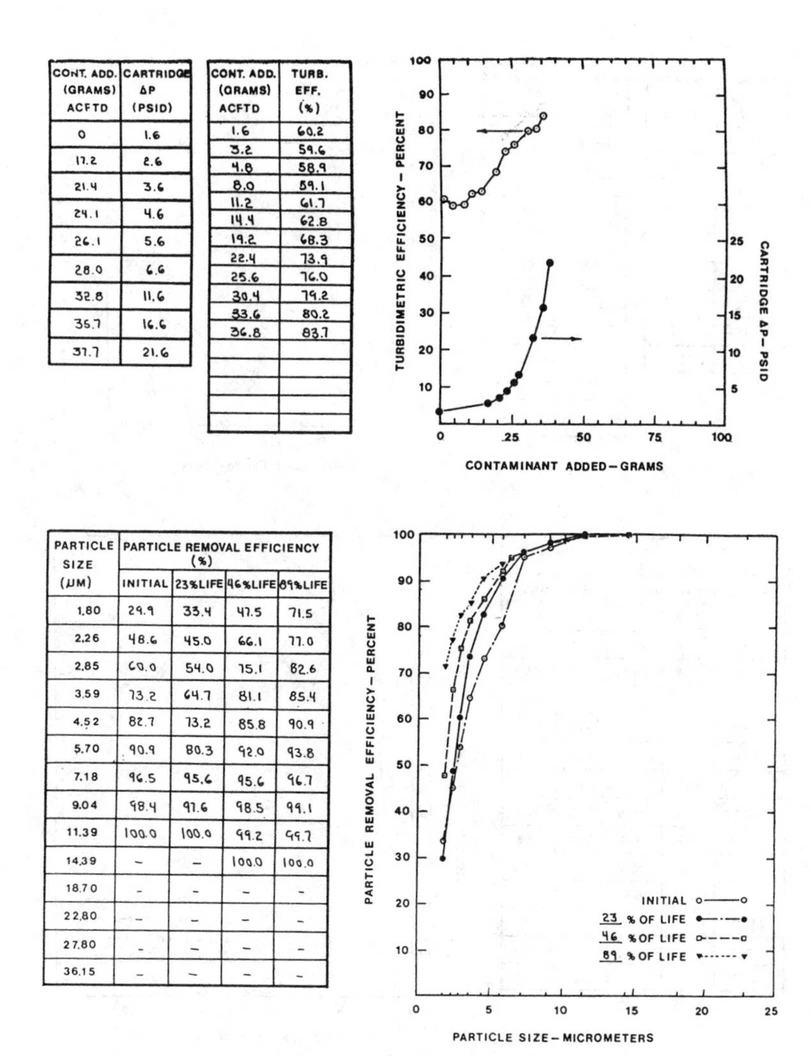

CONT. ADD. (GRAMS) ACFTD	CARTRIDGE ΔP (PSID)
0	1.6
17.2	2.6
21.4	3.6
24.1	4.6
26.1	5.6
28.0	6.6
32.8	11.6
35.7	16.6
37.7	21.6

CONT. ADD. (GRAMS) ACFTD	TURB. EFF. (%)
1.6	60.2
3.2	59.6
4.8	58.9
8.0	59.1
11.2	61.7
14.4	62.8
19.2	68.3
22.4	73.9
25.6	76.0
30.4	79.2
33.6	80.2
36.8	83.7

PARTICLE SIZE (μM)	PARTICLE REMOVAL EFFICIENCY (%) INITIAL	23%LIFE	46%LIFE	89%LIFE
1.80	29.9	33.4	47.5	71.5
2.26	48.6	45.0	66.1	77.0
2.85	60.0	54.0	75.1	82.6
3.59	73.2	64.7	81.1	85.4
4.52	82.7	73.2	85.8	90.9
5.70	90.9	80.3	92.0	93.8
7.18	96.5	95.6	95.6	96.7
9.04	98.4	97.6	98.5	99.1
11.39	100.0	100.0	99.2	99.7
14.39	–	–	100.0	100.0
18.70	–	–	–	–
22.80	–	–	–	–
27.80	–	–	–	–
36.15	–	–	–	–

FIGURE 14 - FILTRATION CHARACTERISTICS - MICROWYND D-CCPY FILTER CARTRIDGE

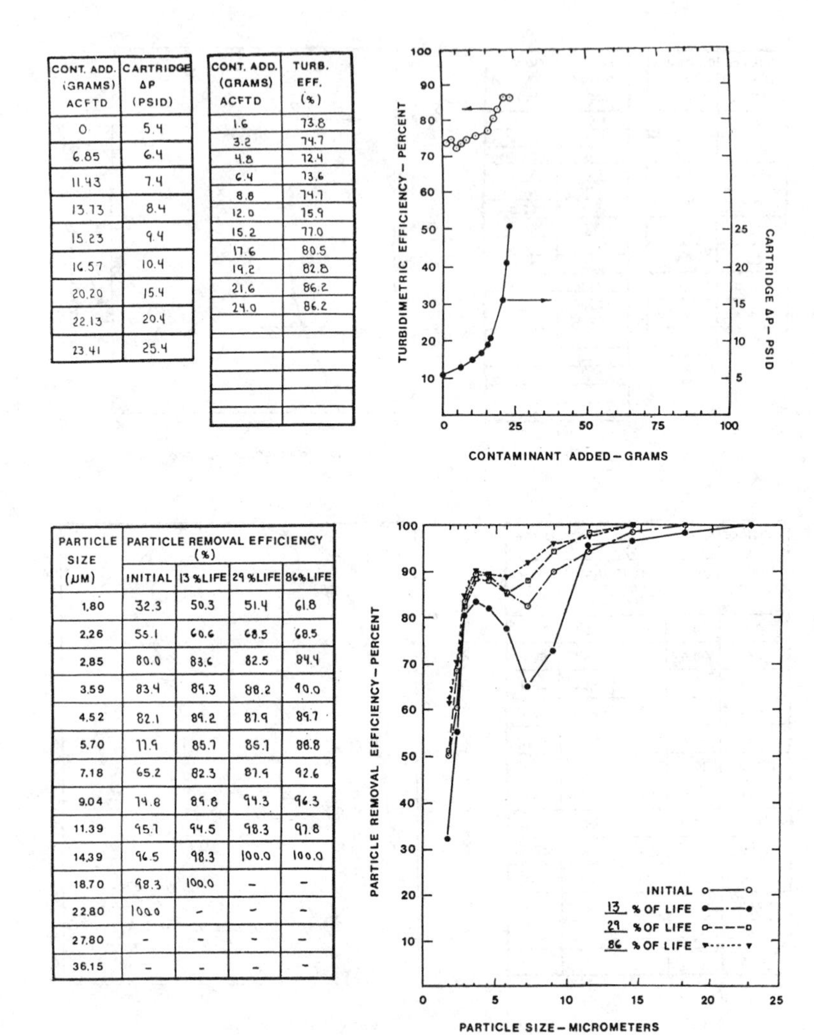

CONT. ADD. (GRAMS) ACFTD	CARTRIDGE ΔP (PSID)
0	5.4
6.85	6.4
11.43	7.4
13.73	8.4
15.23	9.4
16.57	10.4
20.20	15.4
22.13	20.4
23.41	25.4

CONT. ADD. (GRAMS) ACFTD	TURB. EFF. (%)
1.6	73.8
3.2	74.7
4.8	72.4
6.4	73.6
8.8	74.7
12.0	75.9
15.2	77.0
17.6	80.5
19.2	82.8
21.6	86.2
24.0	86.2

PARTICLE SIZE (μM)	PARTICLE REMOVAL EFFICIENCY (%)			
	INITIAL	13 %LIFE	29 %LIFE	86%LIFE
1.80	32.3	50.3	51.4	61.8
2.26	55.1	60.6	68.5	68.5
2.85	80.0	83.6	82.5	84.4
3.59	83.4	89.3	88.2	90.0
4.52	82.1	89.2	87.9	89.7
5.70	77.9	85.7	85.7	88.8
7.18	65.2	82.3	87.9	92.6
9.04	74.8	89.8	94.3	96.3
11.39	95.7	94.5	98.3	97.8
14.39	96.5	98.3	100.0	100.0
18.70	98.3	100.0	-	-
22.80	100.0	-	-	-
27.80	-	-	-	-
36.15	-	-	-	-

FIGURE 15 – FILTRATION CHARACTERISTICS – CONVENTIONAL WOUND FILTER CARTRIDGE – COTTON 39 WIND

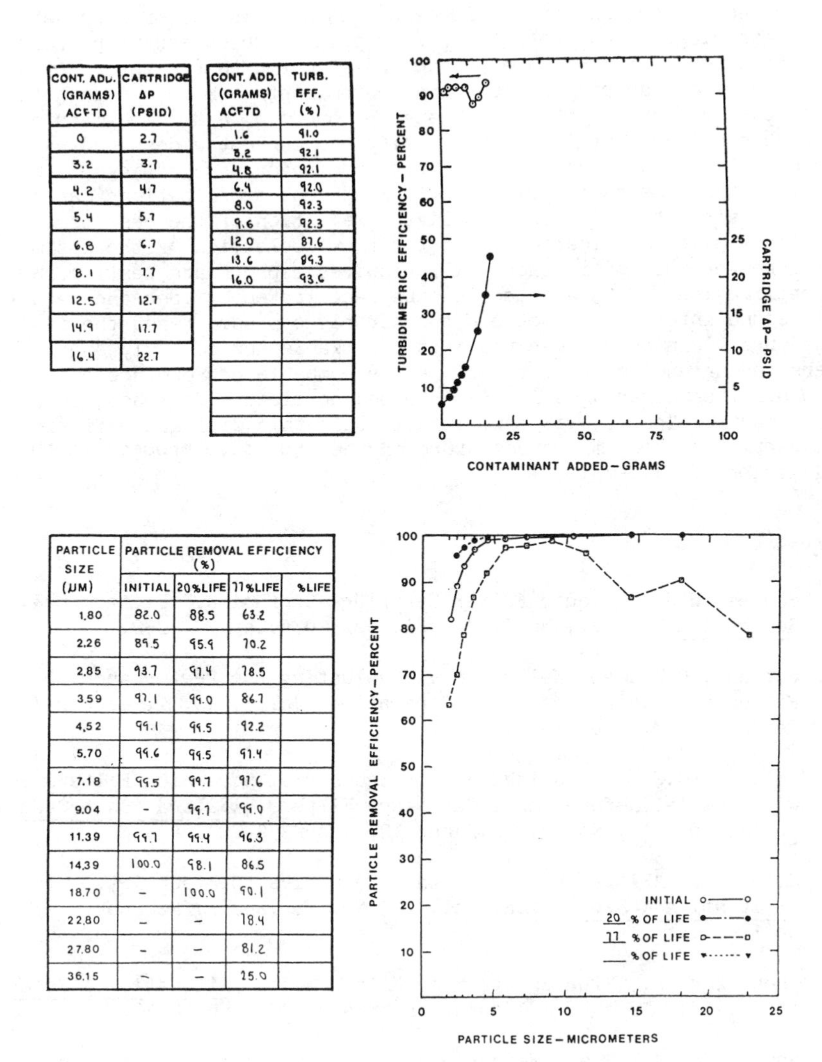

CONT. ADD. (GRAMS) ACFTD	CARTRIDGE ΔP (PSID)
0	2.7
3.2	3.7
4.2	4.7
5.4	5.7
6.8	6.7
8.1	7.7
12.5	12.7
14.9	17.7
16.4	22.7

CONT. ADD. (GRAMS) ACFTD	TURB. EFF. (%)
1.6	91.0
3.2	92.1
4.8	92.1
6.4	92.0
8.0	92.3
9.6	92.3
12.0	87.6
13.6	89.3
16.0	93.6

PARTICLE SIZE (μM)	PARTICLE REMOVAL EFFICIENCY (%)			
	INITIAL	20%LIFE	77%LIFE	%LIFE
1.80	82.0	88.5	63.2	
2.26	89.5	95.9	70.2	
2.85	93.7	97.4	78.5	
3.59	97.1	99.0	86.7	
4.52	99.1	99.5	92.2	
5.70	99.6	99.5	97.4	
7.18	99.5	99.7	97.6	
9.04	-	99.7	99.0	
11.39	99.7	99.4	96.3	
14.39	100.0	98.1	86.5	
18.70	-	100.0	90.1	
22.80	-	-	78.4	
27.80	-	-	81.2	
36.15	-	-	25.0	

FIGURE 16 – FILTRATION CHARACTERISTICS – RESIN BONDED GLASS FIBRE FILTER CARTRIDGE

CONCLUSIONS

In this paper, we have attempted to present a useful approach to defining and describing the performance characteristics of disposable filter cartridges. Our results show complex, time dependent performance characteristics that are unique and specific to the interaction between a given cartridge, fluid and contaminant. We have seen nothing in our evaluation of a broad range of our own, and competitor's disposable filter cartridges, to justify the current widespread use of simplistic, single valued, single point performance ratings. Sufficiently comprehensive cartridge performance characterizations can provide a sufficient basis for intelligent selection of the best candidates for a given application - but the final selection of the optimum cartridge for the application still requires appropriate testing in the application. This approach can be facilitated by open and technically sound inter-change between the cartridge producer and the intended user. In turn, this requires a workable and commonly shared understanding and language to define the probable performance of various filter cartridges in the intended application. I hope that the concept of "personality" will facilitate this dialogue reminding both parties of the imprecise nature of the inductive process in the application of filter cartridges.

REFERENCE:

(1) Soules, W.J., "Liquid Filter Cartridge Confusion," Filtation & Separation Vol. II, No.4, July/August 1974, pp 391-392.

(2) Purchas, D.B., and Wells R.M., "Evaluating the Performance of Filter Media," Filtration & Separation, Vol.8, No. 1, January/ February 1971, pp 47-60.

(3) Pointon, C.W., and Giles, J.W., "Industrial Screening Filters with Special Reference to Cartridge Filters," Filtration & Separation, Vol. 11, No. 3, May/June 1974, pp 259-263.

(4) Cole, F.W. "Filter Ratings - An Alternative to Black Art, "Filtration & Separation, Vol. 12, No. 1, January/February 1975 pp 17-22.

(5) Shoemaker, W., "The Spectrum of Filter Media" Filtration & Separation, Vol. 12, No. 1, January/February 1975, pp 62-68.

(6) Sandstedt, H.N., and Weisenberger J.J., "Cartridge Filter Performance and Micron Rating", Filtration & Separation Vol.22, No. 2, March/April 1985, pp 101-105.

(7) Simms, J.R. "Industrial Turbidity Measurement" , ISA Report 71-728.

(8) Knight, R.A., and Ostreicher, E.A., "Measuring the Electrokinetic Properties of Charged Filter Media, Filtration & Separation, Vol. 18, No. 1, January/February 1981, pp 30-36.

(9) Johnston, P.R., "Turbidity of Streams Around Filters and Particle Size Distributions in Those Streams", Commercial Filters Division, The Carborun Company, Lebanon, Indiana, July, 1975.

(10) Johnston, P.R., and Schmitz, J.E.", A New and Recommended Way to View the Test Performance of Cartridge Filter", Filtration & Separation, Vol. 11, No.6 November/ December 1974, pp581-585.

(11) Fitch, E.C. Encyclopedia of Fluid Contamination Control Hemisphere Publishing, Washington, D.C., 1978.

(12) Hong, I.T., "The Beta Prime - A New Advanced Filtration Theory", Filtration & Separation, Vol. 22, No. 4, July/August 1985, pp. 225-228.

(13) Kinsman, S. "Instrumentation for Filtration Tests", Filtration & Separation, Vol.12, No. 4, July/August 1975, pp. 376-378.

(14) Lieberman, A., "Automatic Particle Measurement Systems and liquid Handling Techniques for Clean Liquids" Journal of Testing and Evaluation, Vo. 3, No. 5, September 1975, pp. 398-403.

(15) Johnston, P.R. and Swanson, R., "A Correlation Between the Results of Different Instruments Used to Determine the Particle Size Distribution in AC Fine Test Dust", Powder Technology, Vol. 32, 1985, pp. 119-124.

(16) ASTM F661-80 "Standard Practice For Particle Count And Size Distribution Measurement In Batch Samples For Filter Evaluation Using An Optical Particle Counter", American Society for Testing and Materials, 1980.

(17) ASTM F662-80 "Standard Method For Measurement And Size Distribution In Batch Samples For Filter Evaluation Using An Electrical Resistance Particle Counter" American Society for Testing and Materials, 1980.

(18) Karuhn, R.F., and Berg, R.H., Practical Aspects of Electronzone Size Analysis Particle Data Laboratories, Ltd., Elmhurst, Illinois, 1982.

(19) Lloyd, P.J., and Ward, A.S., "Filtration Applications of Particle Characterization" Filtration & Separation, Vol. 12, No. 3, May/ June 1975, pp. 246-253.

(20) Ostreicher. E.A., "Electrokinetics in Liquid Filtration" Presented at the Hersey Conference on "Fibres, Filtration and Electrostatics", Hersey, Pennsylvania, August 19-21, 1985.

Anthony J. Caronia, Richard L. McNeil, and Kevin J. Rucinski

FILTER MEDIA CHARACTERIZATION BY MERCURY INTRUSION

REFERENCE: Caronia, A. J., McNeil, R. L., and Rucinski, K. J., "Filter Media Characterization by Mercury Intrusion," Fluid Filtration: Liquid, Volume II, ASTM STP 975, P. R. Johnston and H. G. Schroeder, Eds., American Society for Testing and Materials, Philadelphia, 1986

ABSTRACT: A study was made to determine if mercury intrusion could be used as a test method to characterize filter media. Mercury intrusion describes pore structure and void volume of porous materials.

Mercury intrusion results were compared to actual dirt removal tests on flat sheets of filter media by the multi-pass test method. The results, specifically capacity and efficiency, were used to establish the ability of intrusion to rank filter media by pore structure characteristics. Initial results indicate that mercury intrusion can provide rankings of filter media with greater accuracy and significantly less "mystery" than that associated with established media physical properties measurements.

INTRODUCTION:

Historically, techniques for characterizing filter media have involved physical prperties measurements, such as partial flows, max pore and permeability, flat sheet performance tests and/or product testing. Physical properties measurements are limited in their ability to characterize performance because they are a measure of the end effect of pore structure. Medias with different pore structures

A. J. Caronia, Engineering Manager, R. L. McNeil, Project Engineer, K. J. Rucinski, Project Engineer, Allied Aftermarket Division, Fram, Bendix, Autolite, 55 Pawtucket Avenue, East Providence, RI 02916

could have similar physicals but totally different performance. Flat sheet performance tests, while providing a measure of filtration performance, can be time-consuming (in the event particle counting is required), and have not been fully correlated to other product performance tests. Product performance testing is both time-consuming and expensive because of the associated fabrication building time and the actual product testing.

Mercury intrusion takes a "three" dimensional look at materials, providing measures of void volume, pore structure and (calculated) fiber surface area. The test is quick, providing results in approximately 20 minutes and, through the use of a computer interface, provides a multitude of ways to manipulate the results to gain further insight into the material under study.

Mercury intrusion is not a new technique for evaluating porous materials. In order to evaluate the filter medias used in this study, a modification was made to the standard approach. Intrusion was made to take place starting at 0 kpa, thus allowing measurement of pores equal to or less than 400 microns in diameter to be measured. The top pressure was limited to 350 kpa measuring pores 8 microns in diameter.

Mercury intrusion measurements are based on the theory of the Washburn equation (eq. 1), which states that pressure (P) is directly proportional to surface tension (γ) and the cosine of the mercury/medium contact angle (θ), and inversely proportional to pore radius (r). The surface tension and contact angle of the mercury are assumed constant so the pressure of intrusion is only related to pore radius. By accurately measuring pressure a calculation can be made to determine the pore radius being intruded at a given time. Additionally, a very accurate measure of the volume intruded (void volume) can be made. Surface area approximations are made by assuming cylindrical pore shapes and using the pore radius measured and the volume intruded to calculate a surface area value. Surface area is an approximation only but a relative one for different materials. (The more accurate technique for surface area measurement is by gas adsorption.)

$$P = \frac{2\gamma \cos\theta}{r} \qquad \text{Eq. (1)}$$

DISCUSSION

Three (3) filter medias were chosen for evaluation compromising a range of filtration from "tight" fuel filter performance to "open" lube oil performance. These three grades were evaluated by the flat sheet multi-pass test method and mercury intrusion to establish if any relationship exists between the two techniques. Concurrently measurements were taken of the physical properties to review the relationship which exists between performance and physicals. The results of the multi-pass flat sheet tests are shown in Table 1.

TABLE 1--Multi-pass Test Results (Ref.4)

Media	A	B	C
Apparent capacity (grams)	0.622	1.131	2.234
Micron Size	Efficiency (%)		
> = 5 um	95.02	24.79	10.82
> = 10 um	99.11	47.56	17.55
> = 15 um	99.15	71.72	28.27
> = 20 um	99.22	89.33	42.66
> = 30 um	99.24	98.56	70.67
> = 40 um	99.11	99.40	87.10

Test Cond. Flow 10 gph Area 28 in^2 End Point 10 psi

Sheet A is the tighest filter media, C is the most open and B is the intermediate sheet. Each result is the average of ten (10) individual test runs. Statistical analysis of the data indicates that these three media are different until efficiencies above 30 microns are reached at which time media A & B have statistically alike efficiencies.

Media physical properties were measured to evaluate their ability to distinguish between these medias. These physical properties are presented in Table 2.

TABLE 2--Media Physical Properties

Media	A	B	C
Basis weight ($g/279M^2$)	58,332	43,681	36,000
Caliper (mm)	0.57	0.46	0.51
Permeability (m^3/min)@ 0.5" H20	0.1	0.4	1.7
Max pore (mm H_2O Bubble Point)	414	218	132

Area Max Pore: 0.82 in^2 Permeability: 5.9in^2

The physical properties also indicate a difference between these three grades of media. Sheet A is the tightest sheet based on measurements of permeability, max pore, and partial flow. Sheet B follows and Sheet C is the most open sheet by these measurements. What the physicals do not give is any information about actual pore structure, i.e. is sheet A made up of a narrow distribution of many pore sizes? Filtration is affected by pore structure and mercury intrusion measures the structure so conclusions can be drawn about structure effects on performance. Before getting involved with the specific structures of these three medias, Fig. 1 is presented to provide an understanding of typical intrusion test results.

INTRUSION OF MEDIA

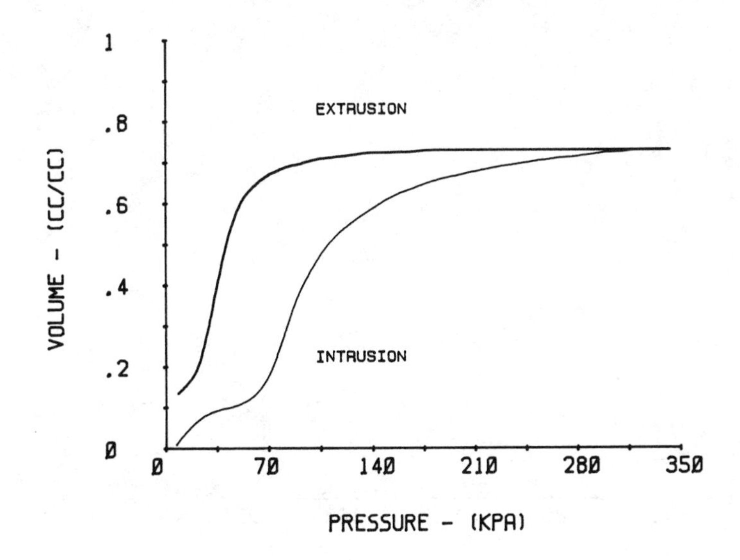

SAMPLE = 17 PAPER VOL=.787 CC

MAX. VOL. = .731 CC/CC

TOTAL SURFACE AREA = 2031 SQ.CM./CC

FIG. 1 Mercury Intrusion of Filter Media

In Fig. 1, the abscissa (x axis) is a measure of intrusion pressure from 0 kpa to 350 kpa. This is the actual pressure range over which intrusion took place. The ordinate (y axis) is a measure of the volume of mercury intruded into the sheet normalized for sample volume, i.e. sample area times caliper. The lower curve is the intrusion curve while the upper curve is the extrusion curve. The intrusion curve is a measure of the pore throat sizes. The extrusion curve is not fully understood, but the hysterisis between the curves is believed to be an indication of pore shape, i.e cylindrical, ink bottle, etc. Other information provided in Fig. 1 is sample number (for records keeping purposes), paper volume (sample area times caliper), maximum volume intruded and total (calculated) surface area.

Another useful variation of the information presented in Fig. 1 is shown in Fig. 2.

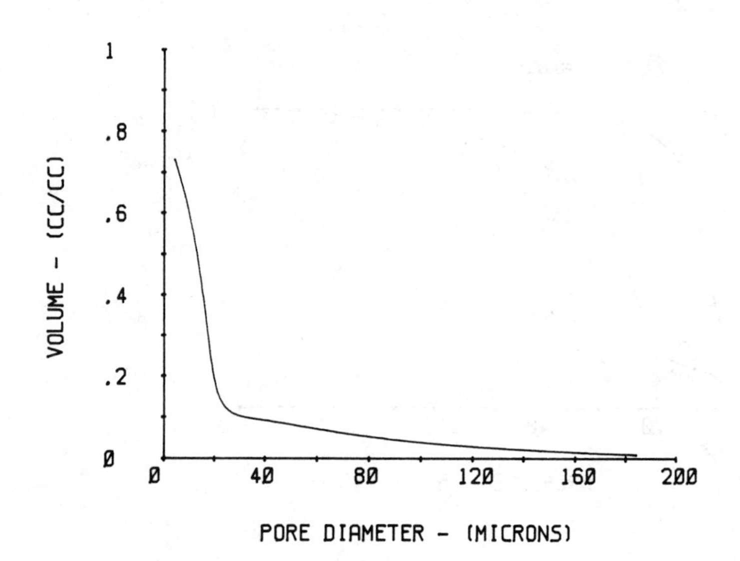

SAMPLE = 17 PAPER VOL=.787 CC

MAX. VOL. = .731 CC/CC

TOTAL SURFACE AREA = 2Ø31 SQ.CM./CC

Fig. 2 Volume vs. Pore Diameter

In Fig. 2 the pressure axis has been mathematically converted to equivalent pore diameters to study the contribution to void volume of each pore size. Note that the majority of void volume for this particular media is contributed by pores in the 5 to 20 micron range, a fairly narrow range of pore sizes.

Referring now to Figs. 3, 4 and 5, these present the data for the three sheets under study.

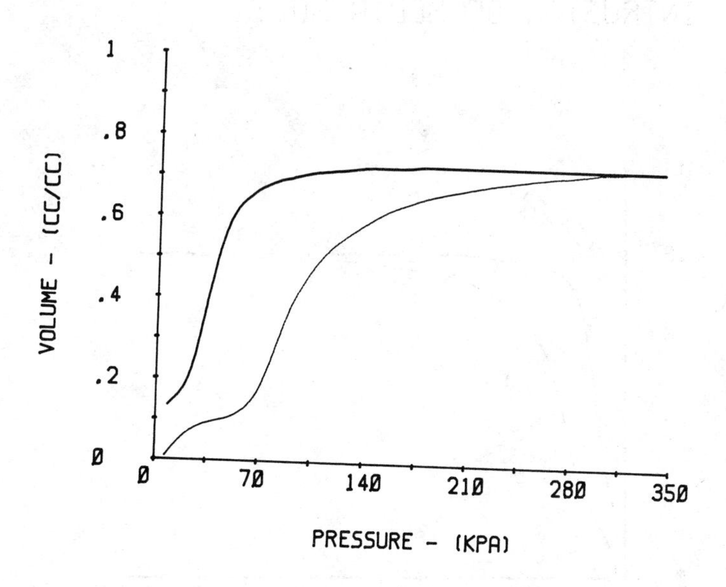

SAMPLE = 17 PAPER VOL= .787 CC

MAX. VOL. = .731 CC/CC

TOTAL SURFACE AREA = 2031 SQ.CM./CC

Fig. 3 Mercury Intrusion of Media Grade A

INTRUSION OF MEDIA NO. B

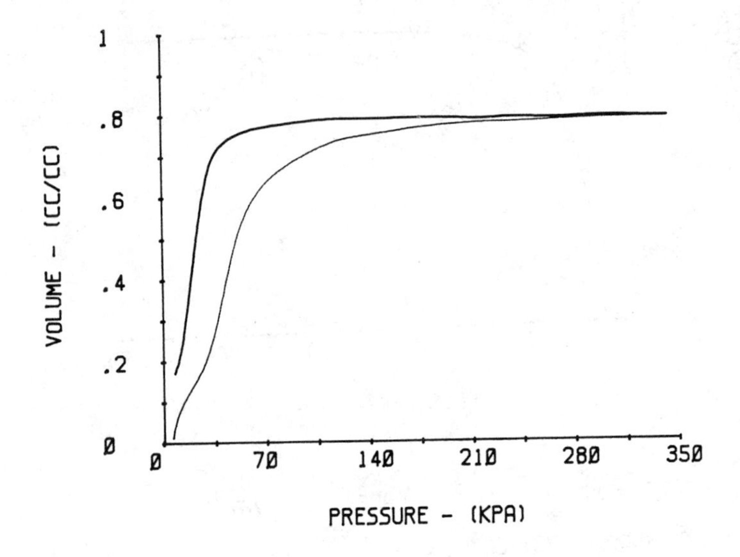

SAMPLE = 12 PAPER VOL= .656 CC

MAX. VOL. = .792 CC/CC

TOTAL SURFACE AREA = 1134 SQ.CM./CC

Fig. 4 Mercury Intrusion of Media Grade B

INTRUSION OF MEDIA NO. C

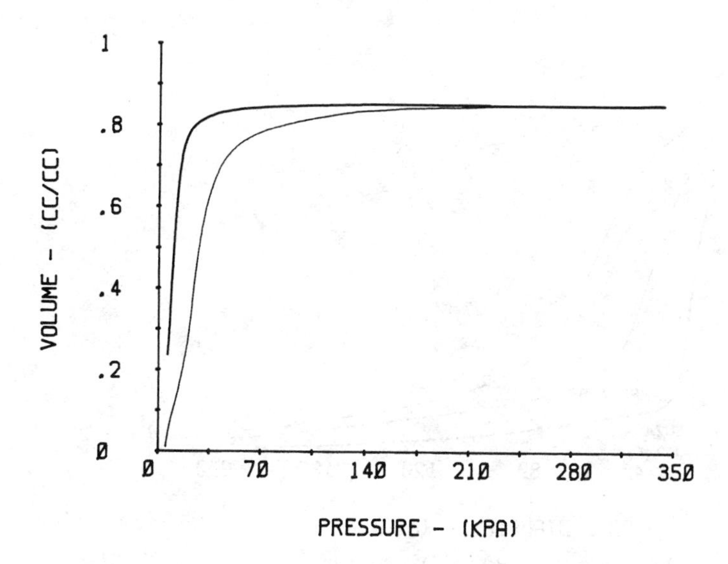

SAMPLE = 15 PAPER VOL=.688 CC

MAX. VOL. = .853 CC/CC

TOTAL SURFACE AREA = 771 SQ.CM./CC

Fig. 5 Mercury Intrusion of Media Grade C

The first notable difference is the appearance of the intrusion curves. Sheet A has the most gradual slope followed by B and then C. This fact by itself is not fully meaningful, but referring to Fig. 6 note that the intrusion of media A is taking place over the narrowest pore range while intrusion takes place over the widest range for sheet C. Also note that sheet A's pore distribution is "shifted" to smaller pore sizes and, in fact, is made up of 5 to 20 micron size pores. Sheet B is made up of 5 to 40 micron size pores and sheet C is made up of 5 to 70 micron size proes. These curves provide some insight into understanding the efficiency numbers given previously.

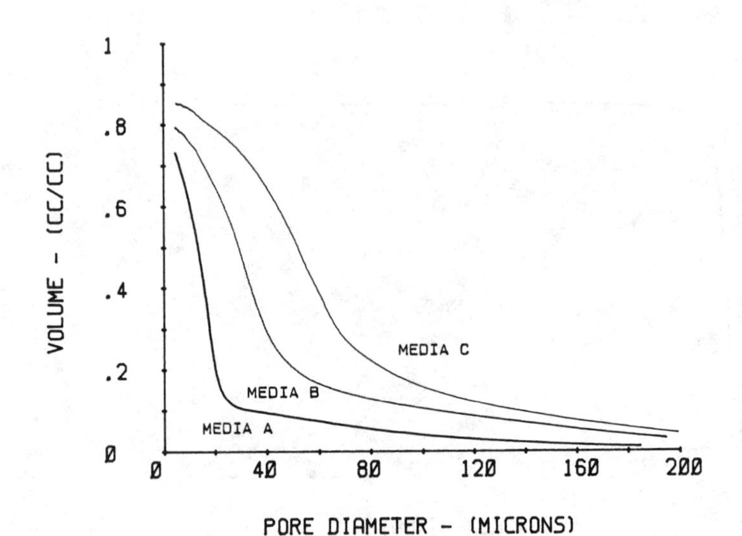

Fig. 6 Volume vs. Pore Diameter

The contaminant used for the multi-pass test was AC fine test dust which has a particle size range from 0 to 80 microns. Looking at the volume vs. pore diameter curve for media A, intrusion indicates the media would be very efficient above 20 microns because there are very few pores above that size for contaminant to pass through. Media B should be very efficient above 40 microns, while media C won't be efficient until contaminant over 70 microns is encountered.

Figure 7 presents a plot of efficiency vs. particle size for the three medias. Media A is exhibiting "perfect" efficiency above 10 microns and slightly lower efficiency below 10 microns. Media B exhibits "perfect" efficiency above 30 microns and media C never reaches 100% at or below 80 microns. The trend of the intrusion data matches the trend of the flat sheet results, i.e. the intrusion plots (Fig. 6) indicate an efficiency "cut-off" above a certain pore size which is not markedly different than the flat sheet results. One potential explanation for the difference is that these efficiency results are equal to or greater than a particular particle size, so efficiencies on smaller

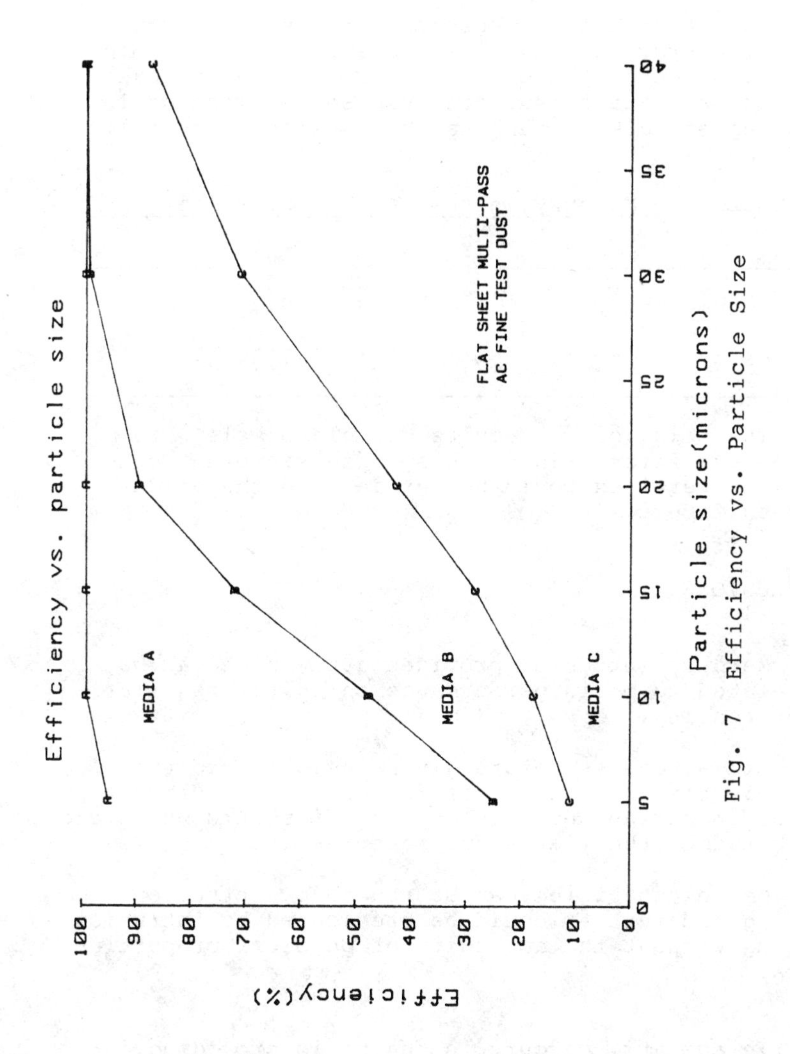

Fig. 7 Efficiency vs. Particle Size

sizes are buoyed up by high efficiencies on larger size particles. Future work will concentrate on the specific efficiencies of given particle sizes, which should shift the flat sheet efficiency results into closer correlation with the mercury intrusion results.

The other piece of useful information mercury intrusion provides is the measurement of void volume. Table 3 presents these results along with the calculated surface areas (surface area shows signs of indicating efficiency ranking, but has not been fully studied).

TABLE 3-- Mercury Intrusion Test Results

Media	A	B	C
Void volume (cc/cc)	0.73	0.79	0.85
Surface area (cm^2/cc)	2031	1134	771

The void volume results correlate well with media capacities. This makes sense since any contaminant retained would be retained in the voids between fibers.

CONCLUSIONS

Mercury intrusion provides data which can be correlated to media performance on multi-pass flat sheet testing.

Media pore structure gives insight into the particle size efficiency of filter media. Mercury intrusion defines efficiency cut-off points which are exhibited on the flat sheet test results.

Media capacities can be ranked by intrusion testing and further could be predicated by intrusion testing without the ambiguity of physical properties tests.

In summary, mercury intrusion is providing insight into the effect on filtration by media pore structure, both for efficiency measurement and capacity measurement. Future work will serve to better quantify the significance of specific media pore sizes and their ability to remove specific contaminant sizes. Also surface area will be studied to establish its effect on media performance.

REFERENCES

[1] Orr, C., "Application of Mercury Penetration to Material Analysis", Powder Technology, 3 (1969/1970), pgs. 117-123.

[2] Coyne, K., O'Brien, R., Conner, W. Curtis, Jr., Rucinski, K., "Filter Morphology and Performance: Porosimetry and Microscopy of Oil Filter Media Compared to Filtration", The Chemical Engineering Journal, publication pending.

[3] Rootare, H. M., "A Review of Mercury Porosimetry, Advanced Experimental. Techniques in Powder Metallurgy", Plenum Press, 1970, pgs. 225-252.

[4] ANSI, B93.31-1973, "Multi-Pass Method for Evaluating the Filtration Performance of a Fine Hydraulic Fluid Power Filter Element," NFPA Std. T3.10.8.8-1973

Peter R. Johnston

THE KOZENY-CARMAN CONSTANT AND THE MOST-PROBABLE, AVERAGE PORE SIZE IN FLUID FILTER MEDIA

REFERENCE: Johnston, P. R., "The Kozeny-Carman Constant and the Most-Probable, Average Pore Size in Fluid Filter Media," Fluid Filtration: Liquid, Volume II, ASTM STP 975, P. R. Johnston and H. G. Schroeder. Eds., American Society for Testing and Materials, Philadelphia, 1986

ABSTRACT: The Kozeny-Carman constant, k, provides an emperical conversion factor relating the permeability of a porous medium to the surface/volume ratio of the building blocks. It does so with an assumption of how the average pore size must change with the packing density or porosity. But, k is not constant--as Carman reports. The present paper offers two reasons why k is not constant: (a) The tortuosity factor (ratio of average pore length to the thickness of the medium) changes with porosity. Carman assumed T is fixed at $\sqrt{2}$ and is a component of k. (b) In the case of a random array of fibers, or sintered particles, the Kozeny method of deducing the average pore size as a function of porosity does not agree with results of probability calculations and of experimental results.

KEYWORDS: Kozeny-Carman constant, pore-size distribution, average pore size, permeability

CONCEPT OF PORE SIZE

A necessary background to the present discussion is a review of the concept of pore size in a random pore-size distribution. Pore size is a statistical concept in which one views a plane in the filter medium perpendicular to the flow of fluid; and, in this view, the medium consist of a stack of such planes, each with some "unit" thickness, and with some finite space between each plane. A single plane contains a random array of irregularly shaped openings of different areas. The linear size of an individual opening refers to the ratio of area to perimeter. For convenience this linear measurement is refered to as the radius; and, for further convience, the opening is considered a circle with unit depth (depth of the plane).

Johnston is Senior Project Engineer, Ametek, Inc., 502 Indiana Ave., Sheboygan, Wisconsin 53081

Where the filter medium consist of a random array of building blocks the pore-size distribution in one of these theoretical planes is identical to the distribution in all the other planes; but, a large pore in one plane does not necessarly occur opposite an equally large pore in the adjacent plane. Thus, where fluid passes through the medium it twists and turns on entering and leaving an individual plane in its quest for passage through the stack of planes. The ultimate path of an average fluid element is, thus, longer than the thickness of the medium by a ratio called the tortuosity factor, T (more below).

A mathematical model of a random pore-size distribution can be stated as follows: Consider the theoretical plane above is marked off in squares where the size of an individual square corresponds to the size of the smallest pore. The probability that an individual square is a pore, or part of a pore, corresponds to the porosity, ϵ , the ratio of void volume to bulk volume. Then in a straight-line walk away from that square of x steps the probability of continuing that walk without leaving the pore is ϵ^x. Thus, a plot of the (relative) number, N, of pores of radius x can be drawn from

$$N_x = \epsilon^x$$

And, with the restraint that the smallest pore has a radius of x = 1.0, the cumulative plot can be drawn from [1]

$$\sum_1^x N_x = \int_1^x \epsilon^x dx = \left[\frac{\epsilon^x}{\ln \epsilon}\right]_1^x \quad (1)$$

Fig. 1 shows a normalized plot of Eq 1 for different values of ϵ.

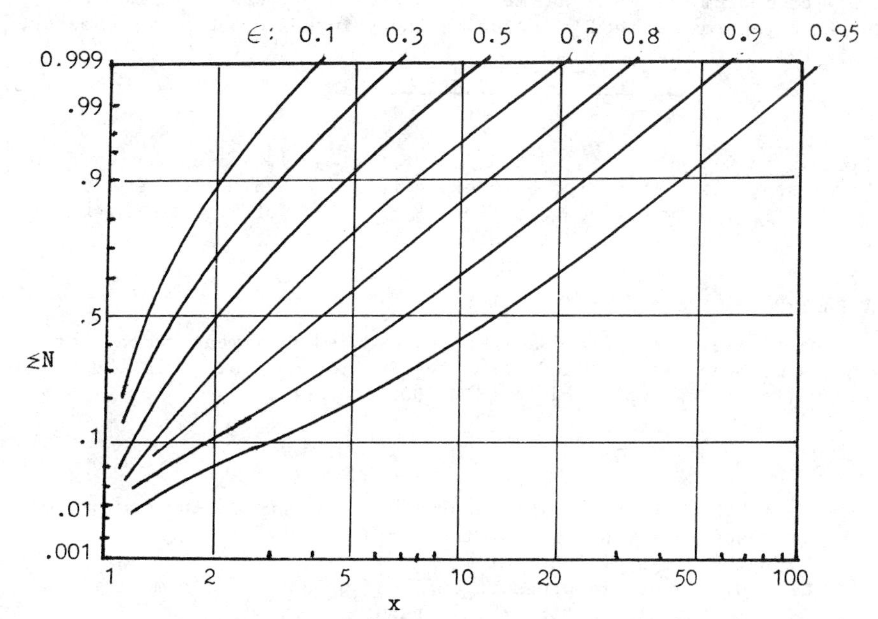

FIG. 1--The most probable pore-size distribution. Plots of normalized values of ΣN in Eq 1 for different values of ϵ.

AVERAGE PORE SIZE

The meaning of an average-size pore in any of the distributions in Fig. 1 rests with what cumulative sum in the distribution one wants to embrace (just as does the meaning of the largest pore). If one embraces the distributions in Fig. 1 as far as the cumulative sum of 0.999 999 99 then the arithmetic average value of x corresponds to the 0.95 mark on the ΣN scale. Alternatively, if one embraces the distribution to only the cumulative sum of 0.999 then the average value of x corresponds to the 0.85 mark on the ΣN scale.

But, the present discussion is more concerned with how the average value of x changes with ϵ on a relative scale. That change is shown in Fig. 2, which is deduced from Fig. 1. Fig. 2 constitutes a basis for the present discussion.

At this point this discussion has been confined to a number distribution of pore sizes. A volume distribution (volume of pores verses pore radius, or diameter) consist of a separate transformation of a distribution in Fig. 1. And, how the viscous flow of fluid is distributed among the pores is shown by another tranformation. Which is to say that for a given porosity the fluid-flow-average pore size is larger than the volume average, which, in turn, is larger than the number average. But, in either kind of distribution the increase in the average pore size with increased porosity follows the ratios in Fig. 2.

THE KOZENY-CARMAN AVERAGE PORE SIZE

On knowing the surface/volume ratio, S/V, of the building blocks in a filter medium, or whem the medium consist of fibers of diameter d_f, the Kozeny-Carman method of deducing the volume-average pore diameter, $\bar{d}$, is [2]

$$\bar{d} = \frac{4\epsilon}{S/V(1-\epsilon)} = \frac{d_f\epsilon}{(1-\epsilon)} \quad (2)$$

which is to say that $\bar{d}$ is a function of $\epsilon/(1-\epsilon)$. For convenience to the present discussion Fig. 3 is drawn from $\bar{d} = 6.57\,\epsilon/(1-\epsilon)$. The curve of Fig. 3 is obviously different from that of Fig. 2--which is the point of the present discussion.

AVERAGE PORE SIZE AND PERMEABILITY

The flow-average pore diameter, $\bar{D}$, deduced from measurements of viscous permeability, B, and porosity, ϵ, is calculated via the Hagen-Poiseuille Law as explained in ASTM F902 and [1]

$$B = \frac{v\,n\,z}{\Delta P} = \frac{(\bar{D})^2\epsilon}{32\,T} \quad (3)$$

where: v = velocity which fluid approaches the face of the medium, m/s
n = absolute viscosity of the fluid, Pa·s
z = thickness of the medium, m
ΔP = difference is pressure on the two faces of the medium, Pa
T = tortuosity factor. With a random array of building blocks within the medium $T = 1/\epsilon$. In a Nuclepore™ membrane T = 1.

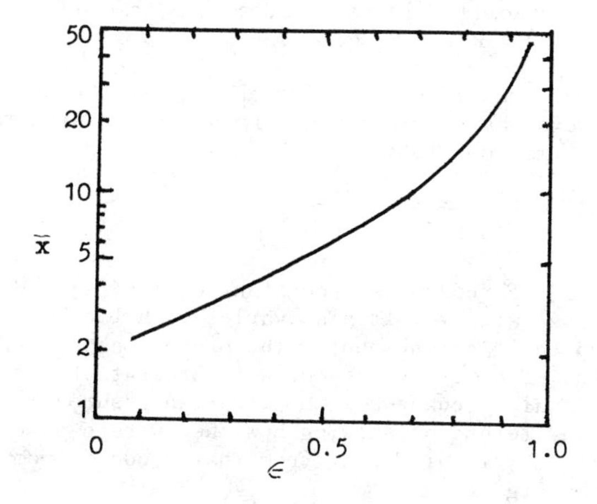

FIG. 2--Average values of x in Fig. 1 as a function of porosity, ϵ. Values correspond to the 0.95 mark on the ΣN scale.

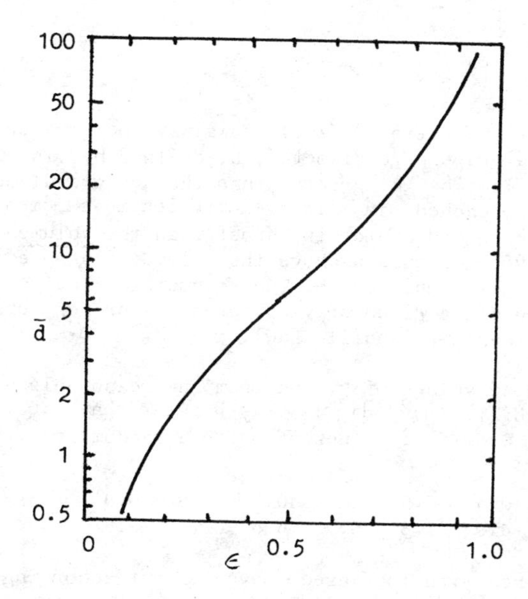

FIG. 3-- Plot of $\bar{d} = 6.57\epsilon/(1 - \epsilon)$, after Kozeny

Before the present practice of employing SI units, previous writers expressed permeability in darcies, rather than units of sq. m. The conversion is 1 darcy = 10^{-12} sq. m.

The Kozeny-Carman school relates permeability to the average pore size $\bar{d}$ of Eq 2 through the emprically determined conversion factor k called the Kozeny-Carman constant

$$B = \frac{(\bar{d})^2 \epsilon}{16\ k} \tag{4}$$

Even though k is refered to as a constant Carman [2] and others [5] show that k varies with ϵ. It also varies with the kinds of building blocks and their arrangements. The present paper offers two reasons why k varies with ϵ: (a) Values of k incorporate the tortuosity factor, T, which certainly changes with ϵ. Carman assumed T is fixed at $\sqrt{2}$; (b) The Kozeny method of showing how the average pore size changes with ϵ, Fig. 3, is different from that deduced from probability calculations, Fig. 2.

Regarding the second point. Consider, from Eqs 3 and 4, the equality

$$\frac{(\bar{D})^2 \epsilon}{32\ T} = \frac{(\bar{d})^2 \epsilon}{16\ k}$$

so that

$$\frac{(\bar{D})^2}{(\bar{d})^2} = \frac{2\ T}{k} \tag{5}$$

for all values of ϵ.

One would expect the ratio $(\bar{D})^2/(\bar{d})^2$ to always be greater than 1.0 simply because the average pore diameter, $\bar{D}$, defined by viscous (Hagen-Poiseuille) flow must be greater than the average diameter, $\bar{d}$, defined by Eq 2--and reached via mercury-intrusion measurements. That is, the flow-average pore diameter is larger than the volume average. In the statistical view discussed above the volume of a pore corresponds to the area with unit depth. The volumetric flow rate of fluid through this area, under a given driving force, is proportional to the square of the area (Hagen-Poiseuille Law).

But, from reported values of k, and from any reasonable assumptions about values of T, $(\bar{D})^2/(\bar{d})^2$ is only greater than 1.0 for small values of ϵ. Consider: Values of k for a random array of fibers are [2]

ϵ--	0.2	0.3	0.4	0.5	0.6	0.7	0.8	0.9
k--	2.7	3.8	4.9	5.8	6.2	6.6	7.2	9.8

Surely T decreases with increased ϵ even if not according to T = $1/\epsilon$. Thus, the ratio $(\bar{D})^2/(\bar{d})^2$ in Eq 5 decreases with increased ϵ. One would expect the ratio to be not only greater than 1.0 but also constant with changes in ϵ. Yet, assuming T = $1/\epsilon$ there is obtained the following ratios:

ϵ	0.2	0.3	0.4	0.5	0.6	0.7	0.8	0.8	0.95
$\frac{2}{\epsilon k} = \frac{(\bar{D})^2}{(\bar{d})^2}$	3.6	1.8	1.0	0.69	0.54	0.38	0.30	0.23	0.20

The reason for this apparent anomaly lies with the understanding of how $\bar{D}$ and $\bar{d}$ change with ϵ, as shown in Figs. 2 & 3. Consider the data in Figs. 2 & 3:

	ϵ	0.2	0.3	0.4	0.5	0.6	0.7	0.8	0.9	0.95
Fig. 2:	$\bar{D}$	2.8	3.5	4.5	5.8	7.2	10	15	30	58
Fig. 3:	$\bar{d}$	1.6	2.8	4.4	6.8	10	15	27	64	124
	$(\bar{D})^2/(\bar{d})^2$	3.1	1.6	1.0	0.73	0.52	0.44	0.31	0.22	0.22

Here the ratios $(\bar{D})^2/(\bar{d})^2$, with changes in ϵ, parallel the ratios above corresponding to $2/\epsilon k$. Thus, the expression of Fig. 3 does not realistically describe the average pore diameter as a function of ϵ. That is, Fig. 2 better describes the average pore size as a function of ϵ than does Fig. 3.

BEDS OF SINTERED PARTICLES

While the above discussion surrounds a bed of fibers the following discussion shows that a bed of sintered particles is like the bed of fibers in that both show the same change in average pore size with a change in porosity. It is important to understand that the pore-size distribution in a bed of particles [3] or spheres [4], as deduced from fluid-intrusion measurements, does not follow the log-normal-type distribution seens in beds of fibers or plastic membranes. But, regardless of that fact, the function of average pore size verses porosity does follow the function in Fig. 2.

Meyer & Smith [6] provide the data from which one may deduce average pore size as a function of porosity in a bed of sintered metal particles. They offer an expression (their Eq 22) which in terms of the present notations states

$$B = \frac{v\,n\,z}{\Delta P} = C\epsilon^{4.1} \qquad (6)$$

for ϵ values in the range 0.2-0.7. In the present notation C is a constant which includes the square of the surface/volume ratio of the particles. Relating Eq 6 to Eq 3

$$C\,\epsilon^{4.1} = \frac{(\bar{D})^2\epsilon}{32\,T}$$

Since Meyer & Smith offer that $T = 1.25/\epsilon^{1.1}$ there evolves the relationship

$$(40\,C)^{0.5}\epsilon = \bar{D} \qquad (7)$$

Letting $(40\,C)^{0.5} = 11.25$

so that at ϵ = 0.4, then $\bar{D}$ = 4.5, one may deduce $\bar{D}$ as a function of ϵ as follows

	ϵ--0.2	0.3	0.4	0.5	0.6	0.7
Fig. 2:	$\bar{D}$--2.8	3.5	4.5	5.8	7.2	10.
Eq 7:	$\bar{D}$--2.25	3.4	4.5	5.6	6.7	7.8
Fig. 3:	$\bar{d}$--1.6	2.8	4.4	6.8	10.	15.

Here $\bar{D}$ from Eq 7 more closely parallels $\bar{D}$ from Fig. 2 than it parallels $\bar{d}$ from Fig. 3. Thus, one must question Eq 2 as a statement of how the average pore diameter changes with porosity for a given mass of building blocks.

REFERENCES

[1] Johnston, P.R., "Fluid Filter Media: Measuring the Average Pore Size and the Pore-Size Distribution, and Correlation with Results of Filtration Tests," Journal of Testing and Evaluation, Vol. 13, No. 4, July 1985, pp. 308-315.

[2] Carman, P.C.,"Fluid Flow Through Granual Beds," Transactions of the Institute of Chemical Engineers (London), Vol. 15, 1937, pp. 150-155; Flow of Gasses Through Porous Media, Academic Press, New York, 1956, Chap. I.

[3] Grace, H.P., "Structure and Performance of Filter Media, Part I," American Institute of Chemical Engineers Journal, Vol. 2, No. 3, Sept. 1956, pp. 307-315.

[4] Haring, R.E. and Greenkorn, R.A., "A Statistical Model of a Porous Medium with Nonuniform Pores," American Institute of Chemical Engineers Journal, Vol. 16, No. 3, 1970, pp. 477-483

[5] Rushton, A. and Griffiths, P., in Chap. 3 of Orr, C., Ed., Filtration Principles and Practice, Part I, Marcel Dekker, New York, 1977

[6] Meyer, B.A. and Smith, D.W., "Flow Through Porous Media: Comparison of Consolidated and Unconsolidated Materials," I&EC Fundamentals, Vol. 24, Aug. 1985, pp. 360-368.

Max S. Willis, Jagadeeshan Raviprakash, and Ismail Tosun

A CONTINUUM THEORY FOR FILTRATION

REFERENCE: Willis, M.S., Raviprakash, J., Tosun, I., "A Continuum Theory for Filtration," Fluid Filtration: Liquid, Volume II, ASTM STP 975, P.R. Johnston and H.G. Schroeder, Eds., American Society for Testing Materials, Philadelphia, 1986

ABSTRACT: Dispersed particulate-fluid system are continuous in each phase but discontinuous throughout the system. If the length scale of the dispersed phase is below the scale of a measuring device, then local measurements reflect average quantities rather than interior phase values. This suggests that a new continuum, at the scale of the local measurement, be established. This paper describes such a continuum and its use in analyzing the permeability function, Darcy's law, and a filtrate rate equation.

KEYWORDS: filtration, Darcy's law, permeability, packed bed, continuum theory

INTRODUCTION

The design of systems in which solid particulates are dispersed in a continuous fluid phase is done predominantly by empirical methods. For example, in liquid filtration new filters are simply copies of existing designs {1}. Such empirical design procedures are expensive because each new design, or modification, must be accompanied by costly experimentation.

Dr. Willis is a professor in the Chemical Engineering Department, University of Akron, Akron, Ohio 44325; Mr. Raviprakash is a graduate student at the India Institute of Science, Bangalore, India; and Dr. Tosun is a professor in the Chemical Engineering Department, Middle East Technical University, Ankara, Turkey.

A reduction in repetitive design experiments and an improved degree of predictability for multiphase processes can be achieved through the development of a multiphase continuum theory. It is the objective of this paper to describe the concepts underlying the development of a such a theory and then to show the results of applying this theory to the process of solid-liquid separation by filtration.

SCALE AND MEASUREMENTS IN MULTIPHASE SYSTEMS

In general, multiphase systems (fogs, bubbly flows, foams, smoke, porous media, filter cakes) are composed of phases in which properties are continuous in each of the phases, but are discontinuous over the entire space of the multiphase system. If the length scale of the dispersed phase is below the scale of a measuring device, such as a pressure probe, then this local measurement reflects an average value rather than a true interior phase value. This suggests then that the fundamental concepts, which hold in each phase at a scale much smaller than that of the local measurement, be averaged to establish a new continuum which is at the same scale as that of the multiphase measurement.

Three scales are used for multiphase systems. The macroscopic scale is that associated with the overall dimensions of the filter cake. The dependent variables at this scale are functions of only time.

A scale which is much smaller that the dimensions of the voids between the particulate matter is termed the microscopic scale. To measure a velocity profile at this scale would require a probe several orders of magnitude smaller than the dimensions of the interstices between the particulate matter. Obviously, such a small velocity probe could not be constructed. At this scale, the dependent variables are functions of position within the interstices and time.

A scale between (mezzo) the microscopic and the macroscopic scale is called the mezzoscopic scale. At this scale, a measuring device, such as a pressure probe, would measure a local average value in a porous medium rather than a true interior phase value. This suggests then that the fundamental concepts, which hold in each phase at the microscopic scale, be averaged to establish a new mezzoscopic continuum which is at the same scale as that of the multiphase measurement. The dependent variables at this scale are functions of position in the porous media and time.

By averaging the governing equations for each phase, information about the details of the conditions at the microscopic scale in each of the individual phases is lost. For example, the shape and surface area of the pores and the velocity profiles within these pores is lost after averaging. Since this detailed information is nearly impossible to obtain, then, from a practical standpoint, not much has been lost.

DARCY'S LAW

Darcy's law is an experimental macroscopic scale correlation which relates the pressure drop across an entire porous medium to the superficial velocity through the porous medium and the permeability of the porous media. Intuitive conversion of Darcy's law from a macroscopic scale correlation to a mezzoscopic scale differential equation violates the fundamental concept of averaging. The concept of averaging is contained in the mean value theorem for integrals {2} which states that a function can be averaged to obtain an average value, but the averaged function cannot be recovered from the average value because there are an infinite number of functions that can give the same average value.

For fluid-particulate systems, the fundamental concept of averaging requires that there exist some mezzoscopic scale equation which, when averaged, will give Darcy's relation. However, the reverse procedure of obtaining a local mezzoscopic equation from the macroscopic scale Darcy's law is not permitted by the mean value theorem and if it is done, erroneous conclusions can be expected. In addition, when Darcy's law is written as a local differential equation, the physical significance of the terms and the equation are not evident. This is why Darcy's equation is sometimes called a constitutive equation that defines the permeability and sometimes it is called a local balance for the conservation of momentum.

CONTINUUM THEORY

Averaging Procedure

The averaging procedure consists of selecting a representative elementary volume (REV) {3-9} that is of the scale of a local measurement. As this REV is moved about in the porous medium, the volume fraction of solid particles and fluid within the REV change, but the volume of the REV remains fixed. The average value of, say, the velocities of all the fluid particles in the REV is assigned to the spatial point that is at the centroid of the REV. In this way, every point in the porous medium has an average fluid velocity and it doesn't matter whether the point is in the fluid phase or the particulate phase. The same procedure is applied to the particulate phase. The net result is that the two phases are represented by two overlapping continuous surfaces over the spatial points of the porous medium.

Multiphase Continuum Equations

The execution of this averaging procedure requires two theorems which convert averages of space and time derivatives to space and time derivatives of average values. Application of these theorems to the micro-scale equations results in a set of mezzo-scale balance equations that contain two new terms in addition to the terms normally present in the micro-scale equations such as the accumulation, convection, generation, and conduction terms. One of these new terms accounts for the effect of phase changes, such as evaporation or condensation, and the other accounts for the fluxes that pass through the surfaces between the two phases. It is this latter term in the motion equation that accounts for exchange of momentum between the fluid and particulate phases.

Constitutive Theory

The constitutive or material property equations for each individual phases at the microscopic level are not necessarily descriptive of material characteristics at the mezzoscopic scale {10}. Thus a fluid can be a Newtonian fluid at the microscopic scale, but it may behave in an entirely different manner at the mezzoscopic scale where the presence of the particulate phase alters the apparent material behavior of the fluid phase.

Since the mezzo-scale continuum suggests constitutive equations that are independent of those at the micro-scale, the postulates of constitutive theory {11-14} can be used to formulate general functional relations for the constitutive relations at the mezzo-scale. It is this set of mezzo-scale continuum equations and constitutive theory that is applied to filtration and flow in a packed bed.

APPLICATION TO FILTRATION

The filter cake is assumed to be isothermal, with no phase changes but the particulates are not stationary and hence the porosity is a function of both position and time {15,16}. The variable porosity reflects the deformation of the particulate phase continuum. Rather than attempt to solve the particulate phase motion equation for this deformation, the porosity can be measured directly using electrical conductivity probes. This experimental determination of the porosity effectively replaces the need to solve the particulate motion equation and leaves only three equations to describe the filtration process. The remaining equations are the continuity equations for solid and fluid phases and the motion equation for the fluid phase.

To determine which terms in the fluid phase motion equation are dominant, characteristic measurable micro-scale quantities are defined and used to convert the dimensioned motion equation to dimensionless form in which the parameters are dimensionless numbers {17}. These dimensionless numbers represent the ratios of effects, where the basis effect is arbitrarily chosen as the viscous force. The following dimensionless parameters appear in the mezzo-level motion equation.

Reynolds' number => ratio of the inertial force to the viscous force

(Reynolds'/Froude) number => ratio of the gravity to the viscous force

Pressure number => ratio of the pressure force to the viscous force

Drag number => ratio of the drag force to the viscous force

The determination of numerical values for these dimensionless numbers requires experimental values for the characteristic quantities and the material properties that appear in these dimensionless numbers. The material properties are mezzo-scale properties and are functions {18} of fluid phase density, the porosity, the temperature, and the invariants of the fluid phase deformation tensor. Since such property functions are not available, an approximation is made in which the dependence on porosity is removed so that single phase property values can be used to evaluate the dimensionless numbers.

The numerical values of the dimensionless parameters obtained from experimental data on the filtration of Newtonian fluids and the flow of non-Newtonian power law fluid through a packed bed indicates that, in both cases, the dominant terms in the mezzo-level fluid motion equation are the terms which represent the interfacial drag force, the pressure force, and the gravity force {19}. Inertial and viscous forces are not important. The absence of the inertial force eliminates the explicit dependence on time and the absence of the viscous force implies that the velocity profile is flat and is not a function of the radial coordinate in a one dimensional axial filtration.

As a result of this dimensional analysis, the continuity equations and fluid phase motion equation that govern variable porosity filtrations for a one-dimensional filtration reduce to the following form

$$\frac{\partial}{\partial t}(\varepsilon_\alpha \rho^\alpha) - \frac{\partial}{\partial z}(\varepsilon_\alpha \rho^\alpha v_z^{\ \alpha}) = 0 \tag{1}$$

$$\frac{\partial}{\partial t}(\varepsilon_\beta \rho^\beta) - \frac{\partial}{\partial z}(\varepsilon_\beta \rho^\beta v_z^{\ \beta}) = 0 \tag{2}$$

$$\varepsilon_\alpha \frac{\partial P^\alpha}{\partial z} - \varepsilon_\alpha \rho^\alpha g_z^{\ \alpha} - \lambda(v_z^{\ \alpha} - v_z^{\ \beta}) = 0 \tag{3}$$

$$\varepsilon_\alpha + \varepsilon_\beta = 1 \tag{4}$$

where

ε = porosity,
ρ = density,
v_z = z-component of the mezzo-level velocity,
P = mezzo-level pressure,
g_z = z-component of gravity,
α = fluid phase,
β = particulate phase,
λ = resistance function for the drag force.

The terms in Equation (3) represent the pressure force, the gravity force, and the interphase drag force. The six unknown functions to be determined from these four equations are

$$\varepsilon_\alpha,\ \varepsilon_\beta,\ \lambda,\ v_z^{\ \alpha},\ v_z^{\ \beta},\ P^\alpha$$

This is a mathematically indeterminate system which can be made determinate by two additional relations that can be obtained from directly measured internal porosity and pressures profiles.

The resistance function λ is property parameter whose functional dependence is obtained from the application of the constitutive theory postulates. The results {19} of applying these postulates indicate that λ must be positive and be a function of the fluid phase density, the porosity, the temperature, and the deformation gradient, that is

$$\lambda = \lambda(\rho^\alpha, \varepsilon_\alpha, T^\alpha, v_{z,z}^{\ \alpha}) \tag{5}$$

where

$$,z = \partial/\partial z.$$

The determination of a rate equation for filtration requires a simple evaluation of the fluid phase motion equation at the exit of the filter cake since the fluid velocity is explicit in Equation (3). The procedure takes advantage of the experimental observation {15-19} that the local porosity is a function of only the fractional cake position ξ. In terms of the independent variables (ξ, t), the continuity conditions and motion equation become

$$\xi\dot{L}\frac{d\varepsilon_\alpha}{d\xi} + \frac{\partial}{\partial\xi}(\varepsilon_\alpha v_z^{\ \alpha}) = 0 \qquad (6)$$

$$\xi\dot{L}\frac{d\varepsilon_\beta}{d\xi} + \frac{\partial}{\partial\xi}(\varepsilon_\beta v_z^{\ \beta}) = 0 \qquad (7)$$

$$\frac{\partial P_o^{\ *}}{\partial\xi} - (\lambda L/P_c^{\ o}\varepsilon_\alpha^{\ 2})\ \{\varepsilon_\alpha(v_z^{\ \alpha} - v_z^{\ \beta}\} = 0 \qquad (8)$$

where

ξ = $z/L(t)$,

$P_o^{\ \alpha}$ = $P^\alpha + \rho g^\alpha z - P_o$,

$P_c^{\ o}$ = $P + \rho^\alpha g^\alpha L - P_o$,

$P_o^{\ *}$ = $P_o^{\ \alpha}/P_c^{\ o}$,

P^o = pressure at the exit,

z^o = axial coordinate measured from the exit.

Evaluation of Equation (8) at the exit of the filter cake with the boundary condition that at $\xi = 0$, $v_z^{\ \beta} = 0$, gives

$$(\varepsilon_{\alpha o}^{\ 2} J_o/\lambda_o)(P_c^{\ o}/L(t)) = \dot{V}/A \qquad (9)$$

where

J_o = $(\partial P_o^{\ \alpha}/\partial\xi)|_{\xi=0}$,

$\dot{V}/A$ = $(\varepsilon_\alpha v_z^{\ \alpha})|_{\xi=0}$,

λ_o = resistance at the septum,

$\varepsilon_{\alpha o}$ = porosity at the septum,

$\dot{V}$ = flow rate,

A = cross sectional area,

$L(t)$ = time dependent length of the filter cake,

and where the septum is that location in the filter medium where the particulate velocity is zero.

The rate equation for filtration, Equation (9), indicates that the conditions at the exit of the filter cake control the filtrate rate and that these exit conditions are the porosity at the filter medium, the dimensionless pressure gradient at the filter medium, and the resistance at the filter medium. During the course of a filtration, particulate matter from the filter cake can enter the filter medium and alter these three factors which, of course, can alter the rate of filtration.

This filtration rate equation applies equally well to liquids which exhibit either Newtonian or non-Newtonian behavior at the micro-scale since the mezzo-scale resistance function λ_o is not a function of the micro-scale fluid deformation but rather the mezzo-scale fluid deformation.

A unique aspect of the motion equation, Equation (8), is that the velocity of both phases is explicit and it is not necessary to integrate this equation to obtain the filtrate rate. A simple evaluation at the exit of cake suffices.

APPLICATION TO PACKED BEDS

In a packed bed, the particulate matter does not move and the porosity is constant. Under these conditions, Equation (2) for the particulate phase is satisfied identically and the continuity condition for the fluid phase, Equation (1), indicates that the fluid phase velocity must be not be a function of the axial coordinate. The motion equation for the fluid phase, Equation (3), becomes

$$\frac{\partial P_o^{\alpha}}{\partial z} - (\lambda/\varepsilon_{\alpha})\, v_z^{\alpha} = 0 \qquad (10)$$

The derivative of the velocity, $v_{z,z}^{\alpha}$, is zero since he velocity is constant. The porosity is also constant and these two effects reduce the functional dependence of the resistance function to

$$\lambda = \lambda(\rho^{\alpha}, \varepsilon_{\alpha}, T^{\alpha}) \qquad (11)$$

Normally, the expression for Darcy's law contains an explicit dependence on the viscosity which is a material transport property at the micro-scale. However, the resistance function depends on the velocity gradient at the mezzo-scale and for flow in a packed bed, no velocity gradients exist at that scale. Consequently the viscosity cannot be introduced as a characteristic of the deformation of the liquid phase. Instead, the viscosity is introduced through the temperature and the Theory of Corresponding States in which the reduced temperature is a unique function of the reduced viscosity

$$T^{\alpha} = T_c\, g(\mu/\mu_c) \qquad (12)$$

where

T_c = critical temperature,
g = a reduced function of the reduced viscosity,
μ_c = critical viscosity,
μ = viscosity.

Upon substitution of Equation (12) into Equation (11), the functional dependence of the resistance function becomes

$$\lambda = \lambda(\rho^{\alpha}, \varepsilon_{\alpha}, T_c, \mu, \mu_c) \qquad (13)$$

It is possible that the viscosity may be factored out of the resistance function and the permeability, K, defined by

$$\lambda = \varepsilon_\alpha^{\ 2}\ \mu\ K^{-1} \tag{14}$$

has the following functional dependence,

$$K = K(\rho^\alpha, \varepsilon_\alpha, T_c, \mu_c) \tag{15}$$

For flow in a packed bed of stationary particulates, the mezzo-level motion equation, Equation (10), becomes

$$\frac{dP_o^{\ \alpha}}{dz} = (\mu/K)(\varepsilon_\alpha\ v_z^{\ \alpha}) \tag{16}$$

where the viscosity, permeability, porosity, and velocity are all constant in the drag term on the right side of the equation and, as a result, the pressure gradient is constant and the pressure profile is linear in the axial direction. It is important to recall that the viscosity in Equation (16) is not characteristic of the micro-scale deformation of the fluid but instead is a thermodynamic quantity that represents the temperature.

When the local pressure gradient is replaced with the pressure drop across the packed bed, Equation (16) becomes Darcy's law

$$(K/\mu)(P_c^{\ o}/L) = \dot{V}/A \tag{17}$$

where

$$\dot{V}/A = \varepsilon_\alpha v_z^{\ \alpha}$$

This development for flow in a packed bed shows that Darcy's law is a macro-scale equation that is restricted by the conditions of a constant porosity and constant fluid velocity. The velocity profile is flat with respect to the radial coordinate and uniform in the axial coordinate. As a result, the fluid at the mezzo-scale is not subject to velocity gradients in either coordinate and, consequently, the rate of deformation of the liquid phase does not appear in the resistance function.

Equation (12), applies to liquids that are either Newtonian or non-Newtonian at the micro-scale even though the viscosity in Equation (12) gives the appearance that Darcy's law is restricted to Newtonian liquids. The viscosity appears in Darcy's law as a thermodynamic property rather than as a transport property.

Although the porosity is constant, it can assume different values when the particulate matter in the packed bed is changed. If the porosity is removed as a variable in the permeability, then there is no distinction between the mezzo-scale and the micro-scale. This contradicts our initial hypothesis that local measurements at the mezzo- and micro-scales are different and hence the permeability must be a function of the porosity.

This development of Darcy's law for packed beds proceeds from the mezzo-scale to the macro-scale and does not violate the averaging concept. Such a development reveals that the two terms in the governing equation, Equation (16), represent the drag force and the pressure force, that the velocity is uniform in both the axial and radial coordinates, that the pressure gradient is constant, and that the pressure profile is linear. Constitutive theory provides the functional dependence of the permeability which is essential to this development because it shows that the permeability is constant in Darcy's law. The reverse procedure of postulating a mezzo-scale differential equation from the macro-scale Darcy's law would not reveal any of the information just discussed.

PACKED BEDS AND FILTRATION

A comparison of the rate equations for a packed bed and filtration indicates that there are significant differences between the two although the equations appear to have the same form.

For a packed bed, Darcy's law

$$(K/\mu)(P_c^{\,o}/L) = \dot{V}/A \tag{17}$$

holds at every point throughout the packed bed and both K and μ are constant. The viscosity represents the temperature rather than the deformation characteristics of the fluid phase. Equation (17) holds for both Newtonian and non-Newtonian liquids.

For a filtration, the filtrate rate equation

$$(\varepsilon_{\alpha o}^{\,2} J_o/\lambda_o)(P_c^{\,o}/L(t)) = \dot{V}/A \tag{9}$$

holds only at the exit of the filter cake. The porosity $\varepsilon_{\alpha o}$, the dimensionless pressure gradient J_o, and the resistance function λ_o are not constant but can change during the course of the filtration due to particulate matter entering the filter medium.

The resistance function in Equation (9) cannot be converted to the permeability function in Equation (17) because λ_o is a function of the mezzo-scale deformation of the liquid phase while the permeability K is not. The viscosity in Equation (17) does not represent mezzo-scale liquid deformation but rather mezzo-scale temperature.

The reason that the rate equations for a packed bed is different from the rate equation for filtration is the presence of the filter medium in filtration but not in a packed bed. The filter medium acts as boundary condition which requires a solution that is different from that for a packed bed. As defined here, Darcy's law and permeability cannot be used to describe a filtration.

CONCLUSIONS

This development of rate equations for a packed bed and filtration demonstrate the importance of scale and the dimensions of measurements in multiphase systems. The existence of different scales combined with the implications of mean value theorem indicate that mechanistic information is only obtained when one proceeds from the micro-scale to the mezzo-scale and finally to the macro-scale. The reverse procedure provides no mechanistic information and is not a sound mathematical procedure.

The effort expended on developing a thermodynamically consistent, mathematically general continuum theory and on applying the postulates of constitutive theory to multiphase systems is justified by the mechanistic information that could not be obtained by either intuitive equations or by extensive experimentation.

Darcy's law and the permeability function apply only to packed beds where the porosity is constant and where the supporting medium for the packed bed offers no impediment to the flow. Darcy's law cannot be used for filtrations where the filter medium alters the boundary conditions at the exit of the filter cake. Even though the viscosity appears in Darcy's law, it is generally applicable to both Newtonian and non-Newtonian liquids.

The filtrate rate equation contains terms which reflect the presence of the filter medium and its effect on the filtrate rate. The penetration of particulate matter into the filter medium causes the septum resistance, porosity, and pressure gradient to change which, in turn, alters the filtrate rate. The resistance function in the filtration equation cannot be converted to the permeability function.

ACKNOWLEDGEMENTS

This work has been supported by the National Science Foundation, Grants ENG-7606224 and CPE-8007523, the Chemical Engineering Department, the Research (Faculty Projects) Committee, and the Office of the Coordinator of Research at the University of Akron.

REFERENCES

{1} Cheremisinoff, N.P., and Azbel, D.S., Liquid Filtration, Ann Arbor Science Publishers, Woburn, 1983.
{2} Courant, R., Differential and Integral Calculus, Interscience Publishers (Wiley), New York, 1937.
{3} Bear, J., Dynamics of Fluids in Porous Media, Elsevier, New York, 1972.
{4} Slattery, J.C., Momentum, Energy, and Mass Transfer in Continuua, McGraw Hill, New York, 1972.
{5} Slattery, J.C., "Multiphase Viscoelastic Flow Through Porous Media," AIChE Journal, Vol. 14, 1968, pp. 50-56.
{6} Whitaker, S., "Advances in Theory of Fluid Motion in Porous Media," Industrial and Engineering Chemistry, Vol. 61, 1969, pp. 14-28.
{7} Drew, D.A., and Segal, L.A., "Averaged Equations for Two-Phase Flows," Studies in Applied Mathematics, Vol. 50, 1978, pp. 205-231.
{8} Gray, W.G., and Lee, P.C.Y., "On the Theorems for Local Volume Averging of Multiphase Systems," International Journal of Multiphase Flow, Vol. 3, pp.333-340.
{9} Hassanizadeh, M., and Gray, W.G., "General Conservation Equations for Multi-phase Systems 1. Averaging Procedure," Advances in Water Resources, Vol. 2, 1979, pp. 131-144.
{10} Hassanizadeh, M., and Gray, W.G., "General Conservation Equations for Multi-phase Systems: 3. Constitutive Theory for Porous Media Flow," Advances in Water Resources, Vol. 3, 1980, pp. 25-40.
{11} Coleman, B.D., and Noll, W., "Thermodynamics of Elastic Materials with Heat Conduction and Viscosity," Archive for Rational Mechanics and Analysis, Vol. 13, 1963, pp. 167-178.
{12} Muller, I., "A Thermodynamic Theory of Mixtures of Fluids," Archive for Rational Mechanics and Analysis Vol. 28, 1968, pp. 1-39.
{13} Muller, I., "Thermodynamics of Mixtures of Fluids," Journal de Mecanique, Vol. 14, 1975, pp. 267-303.
{14} Shapiro, A.M., "An Alternative Formulation for Hydrodynamic Dispersion in Porous Media," in Flow and Transport in Porous Media, A. Verruijt and F.B.J. Barends, eds., A.A. Balkema, Rotterdam, 1981, pp. 203-207.
{15} Willis, M.S., Tosun, I., "A Rigorous Cake Filtration Theory," Chemical Engineering Science, Vol. 35, 1980, pp. 2427-2435.
{16} Willis, M.S., "A Multiphase Theory of Filtration," Progress in Filtration and Separation, 3, ed. R.J. Wakeman, Elsevier, Amsterdam, 1983, pp. 1-56.
{17} Willis, M.S., Bridges, W.G., and Collins, R.M., "A Complete Analysis of Non-Parabolic Filtration Behavior," Chemical Engineering Research and Design (London), Vol. 61, 1983, pp. 96-109.
{18} Willis, M.S., "The Interpretation of Non-Parabolic Filtration Behavior," Encyclopedia of Fluid Mechanics, Vol. 5 ed. N. D. Cheremisinoff, Gulf Publishing Co., West Orange, 1986.

{19} Willis, M.S., Collins, R.M., Bybyk, S., "Theory of Filtration," NATO/ASI Fundamentals of Transport Phenomena in Porous Media, eds. J. Bear, M.Y. Corapcioglu, Martinus Nijhoff Kluwer Academic Publishers Group, Dordrecht, 1986.

Kwan S. Koh

PERMEABILITY AND PORE DIAMETER IN RANDOM FIBROUS MEDIUM IN LIQUID FLOW

REFERENCE: Koh, K.S., "Permeability and Pore Diameter in Random Fibrous Medium in Liquid Flow," Fluid Filtration: Liquid, Volume II, ASTM STP 975, P. R. Johnston and H. G. Schroeder, Eds., American Society for Testing and Materials, Philadelphia, 1986.

ABSTRACT: Permeability of flow through porous medium is of value that could reduce a space metric dimension as a pore diameter. A new look and approach to its measurement is discussed with respect to random fibrous matrix in liquid flow, where either turbulent or laminar flow may apply. An attempt is made here to induce resistive metric properties or parameters for the measurement to be relevant and genuine, for which the spatial characteristics are to be incorporated. Currently, two types of approaches may have been practiced in sizing pore diameter of porous media; one of which employs the Poiseuille's expression for mathematical deduction, and the other being of the bubble point test or mercury intrusion type. In view of operative measurement, new constitutive parameters of porous flow are examined, and a new equation is derived.

KEYWORDS: Permeability, pressure gradient, macroscopic similitude, incompressible fluid, porosity.

In 1846, Poiseuille showed that fluid resistant to flow with respect to absolute viscosity η, in radius r and distance z with small change of r would have differential relation $\frac{\partial}{\partial r}(\eta\frac{\partial v}{\partial r}, 2\pi r \pi z)\delta r$ [1] by double integration we have

$$Q = \frac{\pi R^4 \triangle P}{8 \eta z} \tag{1a}$$

$$D^2 = \frac{32 \eta v z}{\triangle P} \tag{1b}$$

Q(=Av) is volume flow rate, v velocity, radius R(=D/2) at the wall or boundary, $\triangle P$ pressure differential. The state of fluid motion is described in terms of volume and surface forces by Bernoulli and Navier - Stockes equations, and which is linearized and simplified [3,5] in z-direction only ($\nabla P = \triangle P/z$)

$$\frac{\triangle P}{z} = k_1 \eta v + k_2 \rho v^2/2 \tag{2}$$

where the pressure gradient (with ∇ differential operator) and ρ mass density for

Koh is Technical Specialist, Schroeder Brothers Corporation, Nichol Ave., Box 72, McKees Rocks, PA 15136.

steady state motion of incompressible fluid. The utility of this convenient form may be dependent upon the constant K_1 and K_2 that are clearly of a space metric parameter (dimensional factor).

DIMENSIONAL CHARACTERISTICS IN MACROSCOPIC SIMILITUDE

The Kozeny-Carman constants κ_1 and κ_2 are offered with porosity and specific surface S_V [3] to describe fluid flow through porous medium. Equating these with K_1 and K_2 to authenticate the equations (1) and (2) types,

$$K_1 = \kappa_1 \frac{S_V^2 (1-\epsilon)^2}{\epsilon^3}, \quad K_2 = \kappa_2 \frac{S_V(1-\epsilon)}{\epsilon^3} \tag{3}$$

The first term of right hand side equation of (2) is viscosity term with the constant K_1 seems to be identifiable with Poiseuille's equation (1).

Further, author expands the equations in macroscopic similitude as was done in principle for the equation (2) by the process of space average in the differential equation of vectorial expression. Assuming that fibrous matrix is homogeneous and isotropic, we may write in terms of the macroscopic similitude that void area diameter D as a pore diameter and solid part as fiber diameter d_f.

$$D/d_f = \epsilon/(1-\epsilon) \tag{4}$$

The legitimacy may be expressed by writing the definition of porosity, void volume versus total volume of the fibrous mat

$$\epsilon = \frac{N \pi D^2 Z/4}{N \pi D^2 Z/4 + N\pi d_f^2 l_f/4} = \frac{D^2 Z}{D^2 Z + d_f^2 l_f} \tag{5a}$$

N is number of compositions in the specific medium thickness Z that fiber length l_f would equal to the single layer of the medium. Then, $z = l_f$

$$\epsilon = \frac{D^2}{D^2 + d_f^2}, \quad 1-\epsilon = \frac{d_f^2}{D^2 + d_f^2} \tag{5b}$$

which proves the concept or the equation (4) applied. We can also see $S_V \propto (1-\epsilon)$ from the dimensional characteristics. This simplification of two dimensional multi-planar structure is quite essential for the statistical geometry of a fibrous network [2], and for its studies. Finally, if we rewrite only the viscosity term of the equation (2) with the constant k_1, and from the porosity identity,

$$\frac{P}{Z} = \kappa_1 \frac{S_V^2 (1-\epsilon)^2 \eta v}{\epsilon^3} \tag{6a}$$

$$\epsilon^2 = D^2(\epsilon) = \kappa_1 \frac{S_V^2 (1-\epsilon)^2 \eta v' z}{P} \tag{6b}$$

where $v' = v/\epsilon$, the velocity in the medium whose porosity is ϵ. This identifies a similarity with the equation (1b).

VARIABLES AND MEASURANDS

Now, looking into measuring process of the equation (1) or (6b), ΔP test may be viewed as Fig. 1. To be more specific about physical outcomes in the measurements, approaching flow Q_1 (with pressure Pa, velocity v_1) enters medium of porosity ϵ, and experiences entrance contraction (Pc), higher static pressure and velocity head loss (Pm) in the medium, and finally exit enlargement (Px) in the back flow (Pb, v_2). If we spell out these terms that affect system energy in continuity theory, then the pressure differential would be

$$\Delta P = P_1 - P_2 = Pa - Pb + Pm - (Pc + Px) \tag{7}$$

To calculate entrance and exit pressure losses, apply typical velocity head loss with coefficients given as

$$h_c = k_c \frac{v_1{}^2}{2g} \quad \text{where} \quad k_c = \left[1 - \left(\frac{A\epsilon}{A}\right)^2\right]^2 \tag{8a}$$

$$h_x = k_x \frac{(v_1/\epsilon)^2 - v_2{}^2}{2\,g}, \quad \text{where } k_x = 1,\ v_1 = v_2 \tag{8b}$$

For $Pc = \rho\, g\, h_c$, and $Px = \rho\, g\, h_x$,

$$Pc + Px = \frac{\rho\, v^2}{2\,\epsilon^2}\,(1 - 2\,\epsilon^4 + \epsilon^6) \tag{9}$$

Also, for medium total pressure intensity

$$Pm = \frac{Pa}{\epsilon} - \frac{\rho\, v^2}{2\,\epsilon^2} \tag{10}$$

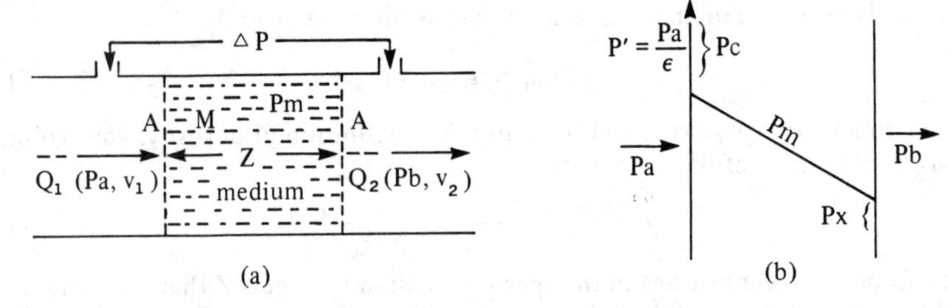

Fig. 1, (a) Medium M is placed in the system of velocity v and pressure P for test. (b) Gradient of the pressure intensity in the medium, ΔP measured for the mathematical deduction is a function of permeability or kinetic energy loss regardless laminar or turbulent.

Letting $Pa - Pb = P_o$, and $P_o = Pa$ in the low back pressure test system as is in the Fig. 1, the above equation (7) can be written as

$$\Delta P = P_o\ (1 + \epsilon) - \frac{\rho v^2}{2\,\epsilon^2}\,\phi_o \tag{11}$$

where $\phi_o = \epsilon^6 - 2\epsilon^4 + 2$

This equation is given for the true pressure differential to be used for the equation (1b) and (6b). Writing $\phi = \phi_o/\epsilon$ for the convenience,

$$D^2 = \frac{32\ z\ v\epsilon\ \eta}{P_o\ (1+\epsilon) - \rho v\phi/2} \tag{12a}$$

$$D^2(\epsilon) = \frac{\kappa_1\ S_v^2\ (1-\epsilon)^2\ z\eta v}{P_o(1+\epsilon) - \rho v\phi/2} \tag{12b}$$

From the macroscopic similitude, the equation (12a) and (12b) are derived for the random fibrous media in a way that Poiseuille's equation applies to measure a pipe diameter or other dimension. The pressure term is adjusted with porosity and inertia term in particular, compared with the original Poiseuille's equation that describes the parabolic flow profile in laminar flow. These modified equations appear to be quite conducive for the theoretical deduction of the pore diameter without complete resistive characteristics (i.e. roughness, adsorptive. etc..).

EXPERIMENTAL METHOD

Test Procedure

As the Fig. 1 shows, simple test circuit is needed to produce an unaltered pressure differential, maintaining equal velocities before and after medium. Fig. 2 and 3 illustrate that, as porosity approaches 1, inertia forces becoming one part of measuring pressure ($\frac{1}{2} \rho v^2 = P$), the equation (12a) returns to normal Poiseuille's equation.

Results and Conclusion

It is reasonable to expect other than laminar flow in porous media where Poiseuille's equation is not applicable in principle. Since no known friction factor or tortuosity associated with flow through media, thus it is reasonable to integrate the porosity with the equations. The nature of porosity alone cannot be defined as resistive metric parameter, hence the integration in constitutive equation may be a heuristic knowledge but deterministically applied here.

Author finds that the modified equation of (12) type suffices where mean flow pore diameter to be measured or the permeability measurement is readily available in most flow tests. Generally, the magnitude of pore diameter produced by the equation shows somewhat smaller than the mean flow size test value.

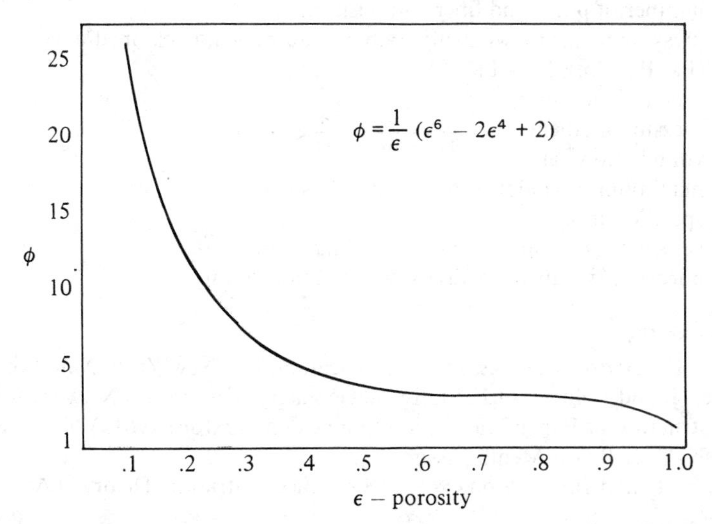

Fig. 2 shows porosity effect in the equations, where the ϕ value derived from overall ϵ effects in the equation, and friction or drag coefficient is not known but the permeability (as a function of ϵ, viscosity, and change of velocities) affects the magnitude of drag force in the ΔP measurement.

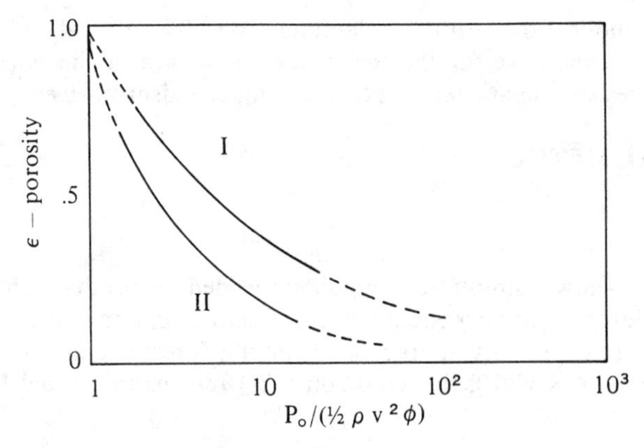

Fig. 3, Usual test ratio of static and inertia forces; I - paper type media, II - open fabric or wire mesh types.

Experimental data for the fluid (depolarized 394 Gage Prod. type)-viscosity 1.7 cP, surface tension 28.5 dyne/cm; in the ranges of z = .15 cm, ϵ = .4 of media; v = .878cm/sec., Po = 22.2 gr/cm^2.

NOMENCLATURE

A : area
D : void or pore diameter; d_f; fiber or solid diameter
g : gravity constant
h_c , h_x : head losses for contraction, exit
k_c , k_x : coefficients of contraction, exit
K_1 , K_2 : constants for equation (2); κ_1, κ_2 : Kozeny constants
N : number of pores and fiber composition
P : pressure intensity, with suffix a,b,m, - approach, back, medium (Pa, Pb, Pm); $P_\circ$ = Pa - Pb
$\triangle$P : pressure differential
∇P : pressure gradient, $\nabla = \frac{\partial}{\partial x} i + \frac{\partial}{\partial y} j + \frac{\partial}{\partial \epsilon} k$
Q : volume flow rate
R : maximum (boundary) radius; r : radius
S_v : specific surface
v : velocity; z: distance, thickness of medium
ϵ : porosity; η: absolute viscosity; ρ: mass density

REFERENCES

[1] Lamb, H., Hydrodynamics, Dover Publication, Inc. New York, New York
[2] Corte, H. and Kallmes, O.J., Statistical Geometry of a Fibrous Network, Formation and Structure of Paper Transaction Symposium, Oxford 1961, Vol. 1, Tech. Sect. British Paper Board Memb. Assoc.
[3] Han, S. T. and Ingermason, W. L., Simplified Filtration Theory, TAPPI Vol. 50 No. 4
[4] Green Jr., L. and Duwez, P., Fluid Flow Through Porous Metals, Journal of Applied Mechanics 1951.
[5] Hubbert, M. K., Field Equations of Liquid Through Porous Solids, International Congress for Applied Mechanics 1946.

Subject Index

Author Index